Unit 4
Biodiversity in Habitats

Copyright © 2020 by Discovery Education, Inc. All rights reserved. No part of this work may be reproduced, distributed, or transmitted in any form or by any means, or stored in a retrieval or database system, without the prior written permission of Discovery Education, Inc.

NGSS is a registered trademark of Achieve. Neither Achieve nor the lead states and partners that developed the Next Generation Science Standards were involved in the production of this product, and do not endorse it.

To obtain permission(s) or for inquiries, submit a request to:
Discovery Education, Inc.
4350 Congress Street, Suite 700
Charlotte, NC 28209
800-323-9084
Education_Info@DiscoveryEd.com

ISBN 13: 978-1-68220-793-2

Printed in the United States of America.

8 9 10 CWD 26 25 24 23   B

© Discovery Education | www.discoveryeducation.com

Acknowledgments

Acknowledgment is given to photographers, artists, and agents for permission to feature their copyrighted material.

Cover and inside cover art: Geoffrey Kuchera / Shutterstock.com

Table of Contents

Unit 4

Letter to the Parent/Guardian v

Biodiversity in Habitats viii

   Get Started: Eco-Friendly Sports 2

Unit Project Preview: Eco-Friendly Outdoor Area 4

Concept 4.1

Living Landscapes 6

   Wonder 8

      Let's Investigate Climbing Perch 10

   Learn 20

   Share 66

Concept 4.2

Plant and Animal Relationships 76

   Wonder 78

      Let's Investigate a Mystery Image 80

   Learn 92

   Share 132

Unit Project

Unit Project: Eco-Friendly Outdoor Area 142

Grade 2 Resources

Bubble Map R1

Safety in the Science Classroom R3

Vocabulary Flash Cards R7

Glossary R13

Index R26

Dear Parent/Guardian,

This year, your student will be using Science Techbook™, a comprehensive science program developed by the educators and designers at Discovery Education and written to the Next Generation Science Standards (NGSS). The NGSS expect students to act and think like scientists and engineers, to ask questions about the world around them, and to solve real-world problems through the application of critical thinking across the domains of science (Life Science, Earth and Space Science, Physical Science).

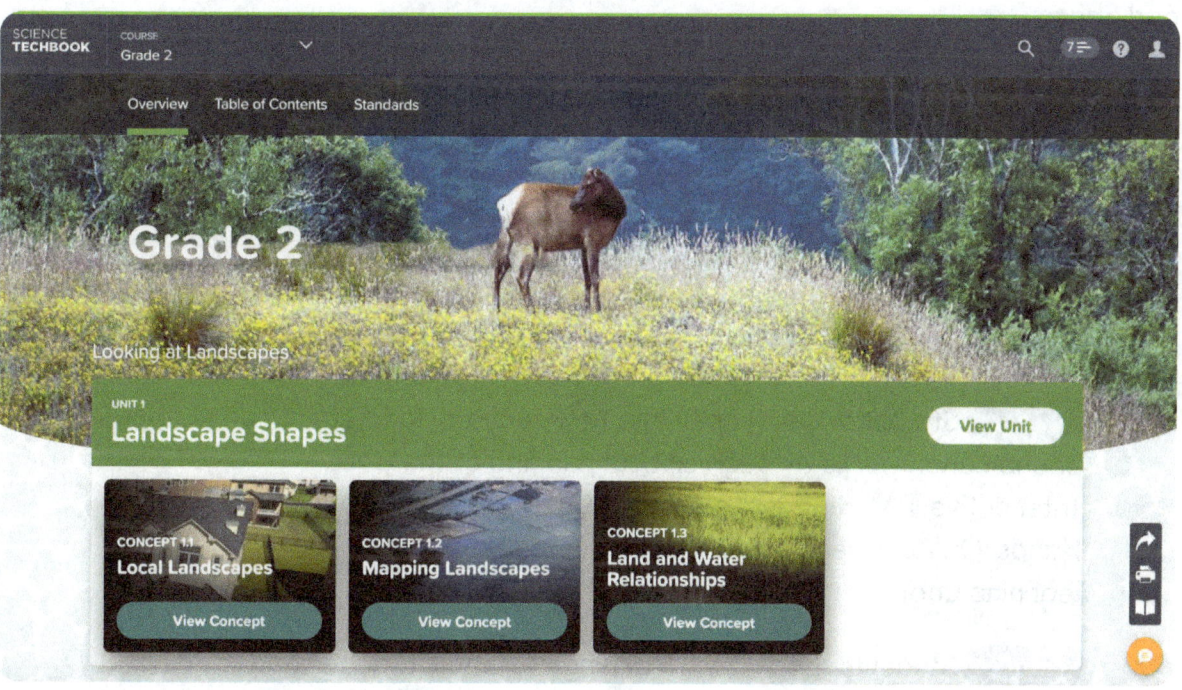

Unit 4: Biodiversity in Habitats

Science Techbook is an innovative program that helps your student master key scientific concepts. Students engage with interactive science materials to analyze and interpret data, think critically, solve problems, and make connections across science disciplines. Science Techbook includes dynamic content, videos, digital tools, Hands-On Activities and labs, and gamelike activities that inspire and motivate scientific learning and curiosity.

You and your child can access the resource by signing in to www.discoveryeducation.com. You can view your child's progress in the course by selecting the Assignment button.

Science Techbook is divided into units, and each unit is divided into concepts. Each concept has three sections: Wonder, Learn, and Share.

Units and Concepts Students begin to consider the connections across fields of science to understand, analyze, and describe real-world phenomena.

Wonder Students activate their prior knowledge of a concept's essential ideas and begin making connections to a real-world phenomenon and the **Can You Explain?** question.

Learn Students dive deeper into how real-world science phenomenon works through critical reading of the Core Interactive Text. Students also build their learning through Hands-On Activities and interactives focused on the learning goals.

Share Students share their learning with their teacher and classmates using evidence they have gathered and analyzed during Learn. Students connect their learning with STEM careers and problem-solving skills.

Within this Student Edition, you'll find QR codes and quick codes that take you and your student to a corresponding section of Science Techbook online. To use the QR codes, you'll need to download a free QR reader. Readers are available for phones, tablets, laptops, desktops, and other devices. Most use the device's camera, but there are some that scan documents that are on your screen.

For resources in Science Techbook, you'll need to sign in with your student's username and password the first time you access a QR code. After that, you won't need to sign in again, unless you log out or remain inactive for too long.

We encourage you to support your student in using the print and online interactive materials in Science Techbook on any device. Together, may you and your student enjoy a fantastic year of science!

Sincerely,

The Discovery Education Science Team

Unit 4: Biodiversity in Habitats

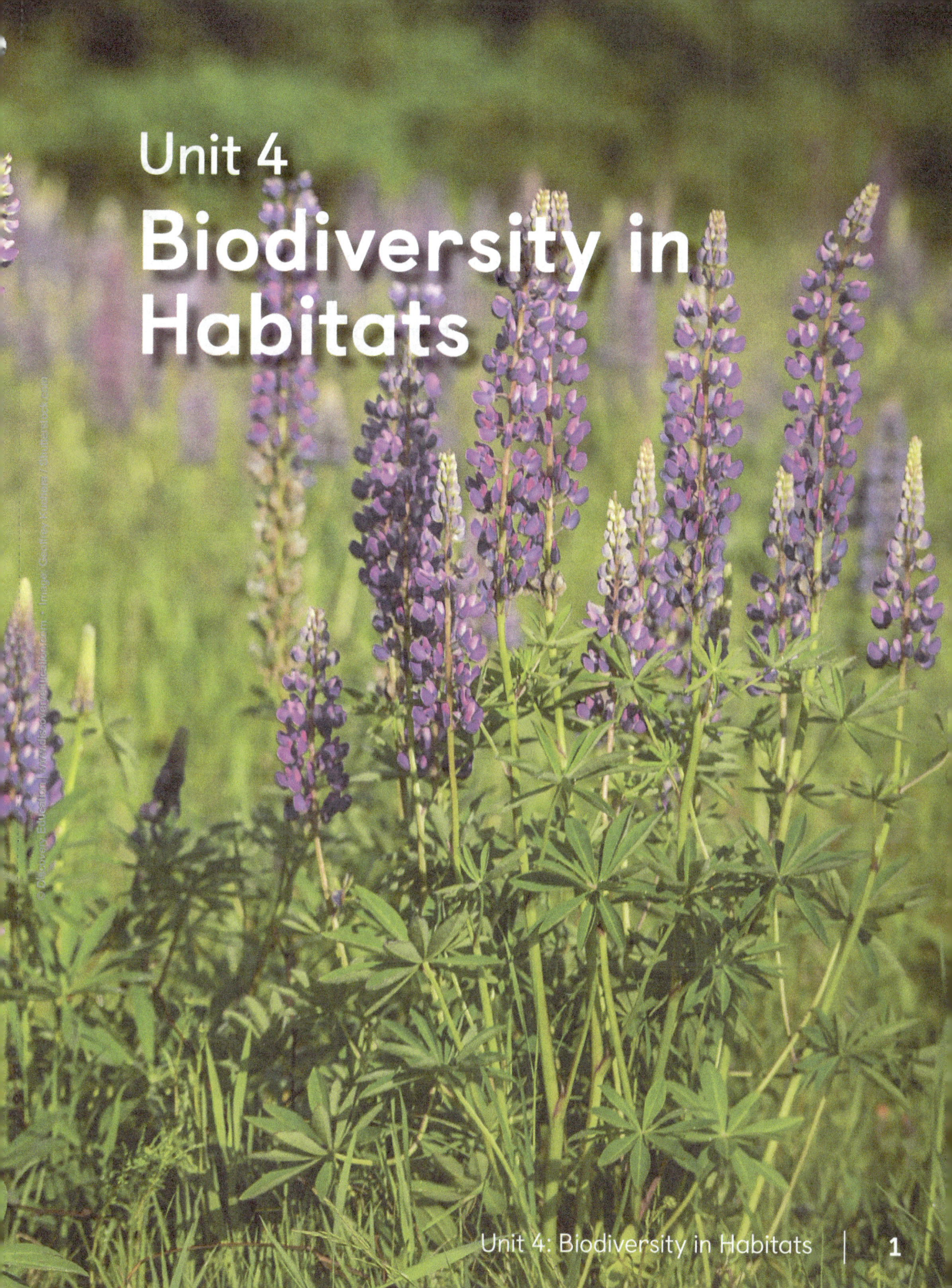

Get Started

Eco-Friendly Sports

Understanding how plants and animals depend on one another will help you think about how changes in a habitat can affect living things. In this unit, you will learn about the variety of plants and animals found in habitats. You will use this information to design an outdoor area for sports that includes eco-friendly features.

Quick Code: us2755s

Eco-Friendly Sports

Think About It

Look at the photograph. **Think** about the following questions:

- How can we determine what plants need to grow?

- How do plants depend on animals?

- How many types of living things live in a place? How can we tell?

Bear Cub Smelling Flowers

Unit 4: Biodiversity in Habitats

Unit Project Preview

Design Solutions Like a Scientist

Quick Code: us2756s

Hands-On Engineering: Eco-Friendly Outdoor Area

In this activity, you will design an eco-friendly outdoor sports area.

Playing Soccer on a Soccer Field

SEP	Asking Questions and Defining Problems
SEP	Developing and Using Models
SEP	Planning and Carrying Out Investigations
SEP	Constructing Explanations and Designing Solutions
CCC	Cause and Effect
CCC	Systems and System Models

Ask Questions About the Problem

You are going to design a new outdoor sports area. You will decide how to build your design so that living things in the area will not be affected. **Write** some questions you can ask to learn more about the problem. **Write** down the answers to the questions as you learn more in this unit.

Unit 4: Biodiversity in Habitats | 5

Student Objectives

By the end of this lesson:

- ☐ I can collect data and count the variety of living things.
- ☐ I can design and do tests to find out what plants need.
- ☐ I can find evidence of how animals obtain energy.
- ☐ I can share how plants and animals depend on each other.

Key Vocabulary

- ☐ biodiversity
- ☐ nutrient
- ☐ organism
- ☐ restore

Quick Code: us2758s

4.1: Living Landscapes | 7

Activity 1
Can You Explain?

How do plants and animals meet their needs in a habitat?

They eat they sleep and they hide and they are born there.

Quick Code:
us2760s

4.1 | **Wonder** How do plants and animals meet their needs in a habitat?

Activity 2
Ask Questions Like a Scientist

Climbing Perch

Look at the image. **Think** about what happened before and after this picture was taken. **Share** with a partner what you think happened before and after this event.

Quick Code: us2761s

Let's Investigate Climbing Perch

Write a story about the picture in six words.

The fish jumped on the rock.

What living and nonliving things do you see in the image?

a fish is living the grass is living the rocks are not.

What questions do you have about the climbing perch fish and its habitat?

Your Questions

4.1: Living Landscapes

 **Read Together**

Living Landscapes

Have you ever seen a fish crawl across a landscape?

Most fish spend their entire lives in water. The climbing perch is not like most fish.

The climbing perch can live out of water for six days. It can crawl across dry land.

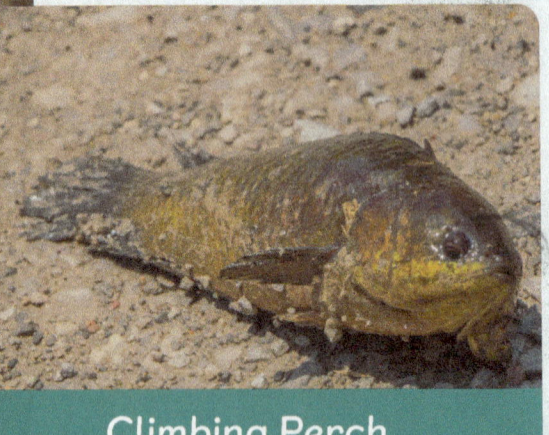
Climbing Perch

The fish can even survive in the mud of a dried-up creek bed for half a year.

Nature can be surprising! What living things can be found near you?

Activity 3
Analyze Like a Scientist

Living Landscapes

Write or **draw** the living things you think are in the schoolyard.

Quick Code: us2762s

4.1: Living Landscapes | 13

4.1 | Wonder — How do plants and animals meet their needs in a habitat?

Activity 4
Observe Like a Scientist

Quick Code: us2763s

The Environment

Watch the video. **Look** at the different environments. Pay attention to the living and nonliving things in each environment.

The Environment

Talk Together

Now, talk together about which things are living and which things are nonliving.

Activity 5
Investigate Like a Scientist

Quick Code:
us2759s

Hands-On Investigation: What Lives Here?

In this activity, take a walk and **look** for living things.

Make a Prediction

You are going to look for living things. **Write** or **draw** your predictions.

Where do you predict you will see living things?

> trees   grass   flowers
> flies   worms   spiders
> bees

SEP Planning and Carrying Out Investigations

CCC Patterns

4.1: Living Landscapes | 15

4.1 | Wonder How do plants and animals meet their needs in a habitat?

What materials do you need? (per group)
- Clipboard
- Paper for clipboard
- Pencils
- Collecting nets (optional)
- Camera (optional)

What Will You Do?

Walk around your schoolyard.

Look for living things in your schoolyard.

Draw pictures of the living things you see.

I see...

As you walk around, **write** anything you wonder about.

I wonder . . .

_____
_____
_____
_____

Think About the Activity

Answer the questions from the patterns you observed.

How were the living things like one another?

_____
_____
_____
_____

How were the living things different from one another?

_____
_____
_____
_____

4.1: Living Landscapes | 17

4.1 | Wonder
How do plants and animals meet their needs in a habitat?

Activity 6
Evaluate Like a Scientist

Quick Code: us2764s

What Do You Already Know About Living Landscapes?

Environmental Factors

Discuss with your group what it is like where you live.
Answer the questions.

Name three words that describe where you live.

big quiet cats

Name three living things that are near where you live.

grass flowers cats

Name three nonliving things that are near where you live.

buildings cars stones.

Getting Our Needs Met

Animals need food, water, sleep, and shelter. **Look** at the pictures of a bear. **Write** which need is met.

	food
	sleep
	water
	shelter

4.1: Living Landscapes

4.1 | Learn How do plants and animals meet their needs in a habitat?

What Living Things Are Found in Habitats?

Activity 7
Observe Like a Scientist

Quick Code: us2765s

Desert Images

Pretend you are a scientist looking for living things in the desert.

Watch the video. **Look** for signs of life. **Make** a mark in the tally box when you see a sign of life.

Desert Images

Your Tally

 **Talk Together**

Now, talk together about the different kinds of life you saw in the video. Share what you saw with your group.

4.1: Living Landscapes

4.1 | Learn How do plants and animals meet their needs in a habitat?

Activity 8
Observe Like a Scientist

Quick Code: us2766s

Joshua Tree National Park

Look at the image. Pay attention to the living things in the park. Then, **answer** the questions.

Walking Through Joshua Tree National Park in California

How many different kinds of plants do you see?
trees   bushes

How many different kinds of animals do you see?
Mamals

22

Activity 9
Analyze Like a Scientist

Quick Code: us2767s

What Living Organisms Are Found in Habitats?

Read about organisms in habitats.

Read Together

What Living Organisms Are Found in Habitats?

A habitat provides an **organism** with food, water, and shelter. Many different organisms can live in one habitat. Plants, animals, and fungi are examples of living things. How do we know how many organisms live in a habitat?

It would be hard for scientists to count all the organisms in each habitat. This would take too long, and they could make a mistake in counting the organisms.

Marking a Sample Grid

4.1: Living Landscapes | 23

Read Together

Scientists use math to help them. They look at small sample areas of a landscape. They count the number of organisms living in the small area to estimate the number of organisms living in the larger landscape.

Where can you look for living things in a landscape?

Look up: You might see birds, butterflies, or tall tree branches.

Look around: You might see grasses, a flower, or a leaping jackrabbit.

Look down: You might see beetles scurrying or a plant growing in the crack of a rock or sidewalk.

Dig a hole: You might find a tortoise burrowed in the sand or an earthworm underground.

Habitats have a great variety of life. You just have to know where to look!

Think about what you have read and the picture you have seen.

Act like a scientist to count butterflies. **Observe** the grid.

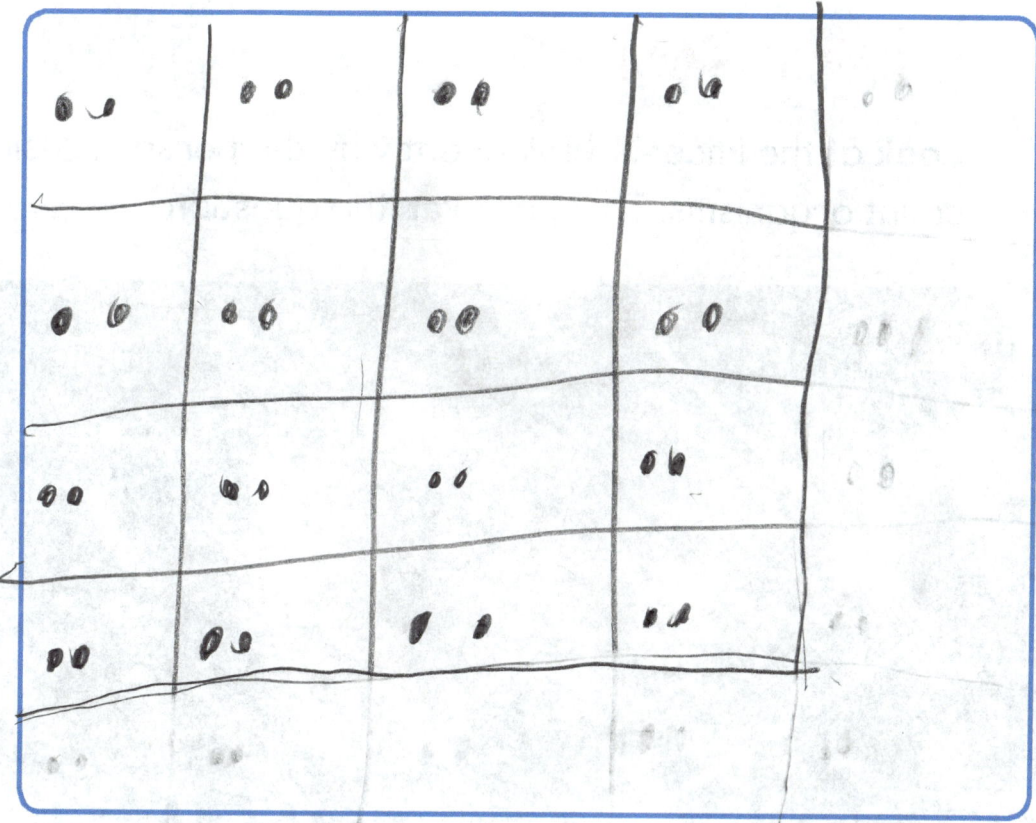

How many butterflies would there be if each cell had two butterflies?

32 butterflies

| SEP | Using Mathematics and Computational Thinking |

4.1: Living Landscapes | 25

4.1 | Learn How do plants and animals meet their needs in a habitat?

Activity 10
Observe Like a Scientist

Quick Code: us2768s

Marking a Sample Grid

Look at the image. **Think** about why the person is using a grid to count organisms. Then, **answer** the question.

Marking a Sample Grid

How do scientists use math to estimate the number of organisms?

They take a sample of a land landscape and they count the organisms.

SEP Using Mathematics and Computational Thinking

Activity 11
Investigate Like a Scientist

Quick Code: us2769s

Hands-On Investigation: Count Them All

In this investigation, you will count and observe the different types of animals in a one square meter area.

Make a Prediction

You are going to look for living things. **Write** your predictions.

How many different kinds of animals do you predict you will see?

I think worms ants and spiders.

How can you set up your area to help you collect your data?

SEP Planning and Carrying Out Investigations

CCC Patterns

4.1: Living Landscapes | 27

4.1 | Learn   How do plants and animals meet their needs in a habitat?

What materials do you need? (per group)

- String
- Measuring tape
- Pencils
- Square (made by string)
- Hand lens
- Parent volunteers (optional but recommended)
- Bamboo skewers
- Clipboard

What Will You Do?

Push the bamboo skewers into the ground to make the square. **Wrap** the string around the bamboo skewers to mark your square area.

Look for animals. **Use** a stick to look under objects.

Count the number of each kind of animal you see.

Describe the area you are looking at. **Sketch** the kinds of animals you find.

Description of Area	Sketch
lots of bees grassy many flowers	

4.1: Living Landscapes | 29

4.1 | Learn — How do plants and animals meet their needs in a habitat?

Write the kind and number of animals you found in the table.

Animal	Eyes?	Legs?	Tally	Count
worms	0	0	////// //// //// //// //// ////	27
ants	2	4	////// //// //// ////	20
beetles	2	6	//// //// //// ////	20
rollypoley	2	8	//	2
snake	2	0	///	3

Think About the Activity

Answer the questions.

Why do you think you found more animals from one category than another?

There were a lot of insects.

In what locations did you find the different animals?

Why would animals need to have different characteristics?

So they would not be the same animal.

4.1 | Learn   How do plants and animals meet their needs in a habitat?

Activity 12
Observe Like a Scientist

Quick Code: us2770s

What Is Biodiversity?

Watch the videos. **Listen** for what the videos tell you about **biodiversity**.

Video

What Is Biodiversity?

Video

Comparing Biodiversity

SEP Obtaining, Evaluating, and Communicating Information

Write a definition of biodiversity.

_____
_____
_____

Think about the data you collected in your schoolyard. **Answer** the question.

Did the area where you counted living things have high or low biodiversity?

_____
_____
_____

What evidence did you use to support your answer?

_____
_____
_____

4.1: Living Landscapes | 33

4.1 | Learn How do plants and animals meet their needs in a habitat?

Activity 13
Evaluate Like a Scientist

Quick Code: us2771s

Comparing Diversity

Different parts of the ocean have different organisms. **Look** at the pictures. Pay attention to what is alike and what is different in different parts of the ocean.

Image 1

Image 2

Image 3

Image 4

SEP Analyzing and Interpreting Data

Write the number of the correct image next to the tally chart.

Tally Chart A
Yellow fish	|
White fish	
Black fish	|
Plant life	|||| |||

Tally Chart B
Yellow fish	|||| ||
White fish	|||| ||||
Black fish	||
Plant life	

Tally Chart C
Yellow fish	||||
White fish	|||
Black fish	|||| ||
Plant life	||

Tally Chart D
Yellow fish	
White fish	
Black fish	|
Plant life	|||| ||||

Answer the questions.
Which image has the most biodiversity?

_____

4.1: Living Landscapes

4.1 | Learn How do plants and animals meet their needs in a habitat?

What Do Plants Need to Live in a Habitat?

Activity 14
Observe Like a Scientist

Quick Code: us2772s

Corn-Covered Landscape

Look at the corn in the picture. **Answer** the questions.

Corn-Covered Landscape

36

What does the corn look like?

The corn looks like baby corn that did not start growing.

Does this corn crop look healthy or unhealthy? Why do you think this?

I think the corn is unhealthy beause the leave are brown and the dirt looks very dry.

4.1: Living Landscapes

4.1 | Learn — How do plants and animals meet their needs in a habitat?

Activity 15
Investigate Like a Scientist

Quick Code: us2773s

Hands-On Investigation: Plant Needs

In this activity, you will learn about the needs of plants. You will learn that plants are living things, so they need things to grow.

You will use plant seedlings to investigate the needs of plants.

Make a Prediction

You are going to find out what plants need.

What does a plant need?

A plant needs soil water and sunlight to survive.

Scientists write questions to investigate. What questions could you write to investigate plant needs?

What if you use prime instad of water to grow a plant.

SEP Planning and Carrying Out Investigations
CCC Cause and Effect

What materials do you need? (per group)

- Seeds, lima beans
- Plastic cup, 5 oz
- Soil, potting
- Light source (such as sunlight or a desk lamp)
- Soap (for handwashing)
- Disposable gloves (per student)
- Craft sticks
- Colored pencils
- Water

What Will You Do?

Place one stick in each plant.

How will you show which plant receives sunlight? How will you show which plant receives water? How will you show which plant receives both sunlight and water?

Draw how you will label the plant needs you are investigating.

4.1: Living Landscapes

4.1 | Learn
How do plants and animals meet their needs in a habitat?

Write the date. **Draw** and **write** about your plants.

Date: _____

Plant 1	Plant 2
Plant 3	Plant 4

Date: _____

Plant 1	Plant 2
Plant 3	Plant 4

Write the date. **Draw** and **write** about your plants.

Date: _____

Plant 1	Plant 2
Plant 3	Plant 4

Date: _____

Plant 1	Plant 2
Plant 3	Plant 4

4.1: Living Landscapes | 41

4.1 | Learn How do plants and animals meet their needs in a habitat?

Write the date. **Draw** and **write** about your plants.

Date: _____

Plant 1	Plant 2
Plant 3	Plant 4

Date: _____

Plant 1	Plant 2
Plant 3	Plant 4

Think About the Activity

Answer the questions.

What did you observe about the plant that did not get any water?

_____

_____

What did you observe about the plant that did not get any light?

_____

_____

Why did you learn about plant needs?

_____

_____

_____

_____

_____

4.1: Living Landscapes | 43

4.1 | Learn How do plants and animals meet their needs in a habitat?

Activity 16
Observe Like a Scientist

Quick Code: us2774s

What Plants Need to Live

Watch the video. Pay attention to how plants grow in different conditions.

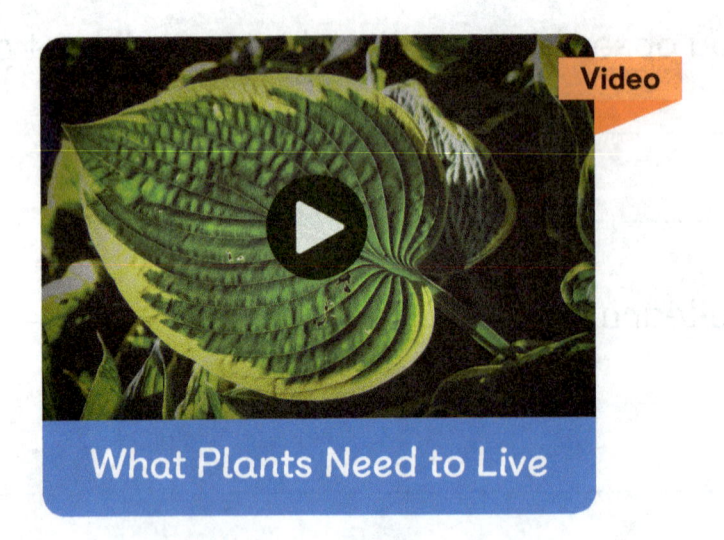

Video

What Plants Need to Live

Talk Together

Now, talk together about what plants need to live.

Try to trick your partner as they guess if you are telling true facts or false facts from the video.

44

Activity 17
Observe Like a Scientist

Quick Code: us2775s

Planting Seeds

Watch the video. **Find** out how you can know where to plant seeds.

Video

Planting Seeds

Talk Together

Now, talk together about where the seeds will grow best.

4.1: Living Landscapes | 45

Read Together

What Do Plants Need to Live in a Habitat?

Plants need water, sunlight, **nutrients**, and space to grow. Plants grow best in locations that meet all these needs. Different plants need different amounts of water, sun, and space. They may also grow best in different kinds of soil.

Schoolyard

Some plants prefer to grow in locations with lots of sun and rich soil. Other plants prefer locations with more shade. There are some plants that grow best in clay soil, while others grow best in sandy soil. What places around your schoolyard are best for growing different kinds of plants?

Activity 18
Analyze Like a Scientist

Quick Code: us2776s

What Do Plants Need to Live in a Habitat?

Think about where you have seen plants grow in your schoolyard.

Look at the Schoolyard image.

How would you describe the sunlight in this schoolyard? **Draw** or **write** your answer.

4.1: Living Landscapes | 47

4.1 | Learn How do plants and animals meet their needs in a habitat?

Activity 19
Think Like a Scientist

Quick Code: us2777s

Mapping Sun and Shade

In this activity, you will make a map of the schoolyard that shows which areas have the most sun and shade. Be sure you wash your hands after touching soil and never put materials into your mouth.

What materials do you need? (per group)
- Small paper observation journals or paper and clipboards
- Pencils
- Small rake or shovel
- Camera (optional)
- Strainers
- Hand lens

SEP Planning and Carrying Out Investigations

CCC Patterns

CCC Systems and System Models

What Will You Do?

Map the schoolyard and show sunny and shady areas.

4.1: Living Landscapes | 49

4.1 | **Learn** How do plants and animals meet their needs in a habitat?

Look, feel, and **smell** the soil in three areas.
Write notes about what you notice about the soil.

Sunny or Shady?	Notes on Soil

Think About the Activity

Analysis and Conclusions

Look at the data you collected about soil in sunny areas. What patterns do you observe?

_____
_____
_____

Look at the data you collected about soil in shady areas. What patterns do you observe?

_____
_____
_____

How are these areas different from each other?

_____
_____
_____
_____
_____

4.1 | Learn How do plants and animals meet their needs in a habitat?

What Do Animals Need to Live in a Habitat?

Activity 20
Observe Like a Scientist

Quick Code: us2778s

Mysterious Mounds

Look at the image.

Mysterious Mounds

Talk Together

Talk together about what may have caused the dirt mounds.

Look at the image. This mole made the mysterious mounds! **Think** about how the mole uses its body.

Mole

Talk Together

Talk together about how the mole's claws help it survive.

4.1: Living Landscapes | 53

Read Together

What Do Animals Need to Live in a Habitat?

Many animals live underground. Prairie dogs, beetles, and earthworms are just a few of the many animals living underground. A mole eats earthworms it finds in the soil.

Have you ever had the chance to observe an earthworm? Sometimes, after it rains, you will see them come up to the surface. Why do they do this?

Living underground provides the earthworm with shelter, water, and dead grass and leaves to eat. The tunnels the earthworm makes bring air to the soil. But when it rains a lot, the soil can be filled with too much water. The earthworm needs air, so if there is too much water, the earthworm will come out of its underground shelter to get air.

Observing an Earthworm

Activity 21
Analyze Like a Scientist

Quick Code: us2779s

What Do Animals Need to Live in a Habitat?

Think about what you have read and the pictures you have seen.

Answer the questions.

What different animals can you find underground?

_____

_____

How do these animals meet their needs? **Write** or **draw** your answer.

4.1: Living Landscapes

4.1 | Learn — How do plants and animals meet their needs in a habitat?

Activity 22
Observe Like a Scientist

Quick Code: us2780s

Shelter

Watch the video. Pay attention to what the worms like about living underground.

Video — Shelter

Talk Together

Now, talk together about what you learned about earthworms and where they live.

Draw and **write** what you learned from the video.

Drawing	Writing

4.1: Living Landscapes | 57

Activity 23
Analyze Like a Scientist

Quick Code: us2781s

Where Are Living Things Found?

Read the passage about living things and habitats. Then, complete the activity that follows.

> Read Together

Where Are Living Things Found?

Look out the window.

What do you see? Living things are all around us. Living things are found all over Earth.

Plants, animals, and people need to live in spaces or places where air, food, and water are available. Living things need habitats that meet their basic needs.

Tree Frog on a Branch

SEP Constructing Explanations and Designing Solutions

58

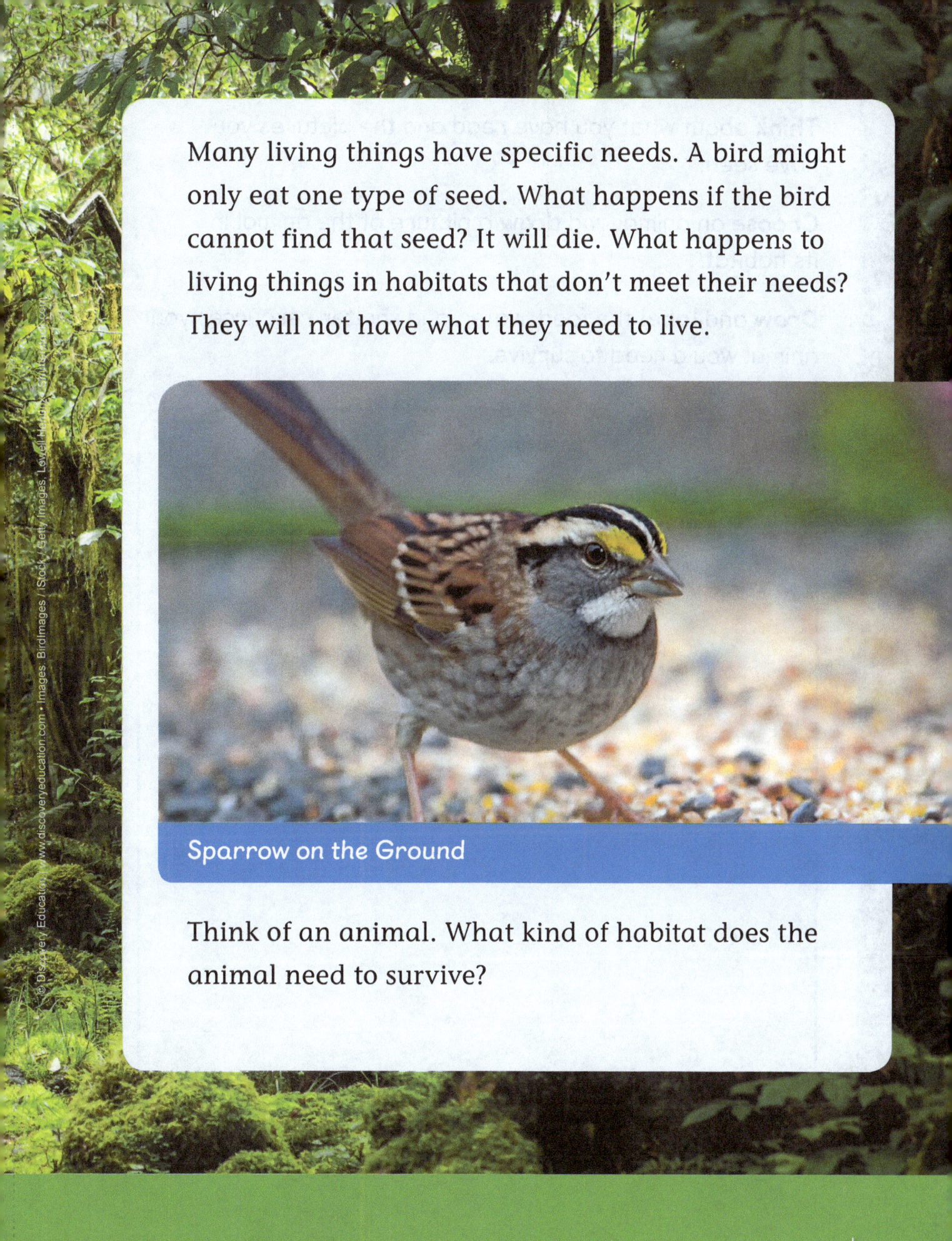

Many living things have specific needs. A bird might only eat one type of seed. What happens if the bird cannot find that seed? It will die. What happens to living things in habitats that don't meet their needs? They will not have what they need to live.

Sparrow on the Ground

Think of an animal. What kind of habitat does the animal need to survive?

4.1: Living Landscapes | 59

Think about what you have read and the pictures you have seen.

Choose an animal and **draw** a picture of the animal in its habitat.

Draw and **label** the food, water, and shelter resources your animal would need to survive.

Your Animal and Its Habitat

Activity 24
Observe Like a Scientist

Quick Code: us2782s

Needs of Living Things

Do the interactive for Life in the Woods. As you visit each forest habitat, **think** about how that habitat has what organisms need. Then, **answer** the question.

Interactive: Needs of Living Things

The fox is able to live in the forest because the habitat meets her survival needs. What three things does the fox get from her habitat that help her to survive?

_____

_____

SEP Developing and Using Models

4.1: Living Landscapes | 61

4.1 | Learn — How do plants and animals meet their needs in a habitat?

Write the organisms that live in each forest habitat. Then **write** the organisms that do not live in that habitat.

Name of Habitat	Live Here	Do Not Live Here
Forest floor		
Upper trees		
Fallen log		
Stream		

Use your table and what you remember from the activity to **answer** the questions.

What three things does a habitat provide for living things?

_____

_____

A sunflower needs sunlight. Why can't it live in the forest floor habitat?

_____

_____

You saw that a forest contains different habitats. Now **think** of the animals and plants that live in or near a freshwater pond. **Describe** three different pond habitats.

_____

_____

_____

_____

4.1: Living Landscapes

4.1 | Learn — How do plants and animals meet their needs in a habitat?

Activity 25
Evaluate Like a Scientist

Quick Code: us2783s

Comparing Habitats

Look at the picture with the ice-covered mountain. **Circle** the living and nonliving things you see in the left circle. **Look** at the picture with the green grass. **Circle** the living and nonliving things you see in the right circle. What living and nonliving things are in both pictures? In the area where the circles overlap, **circle** the living and nonliving things that are common to both habitats.

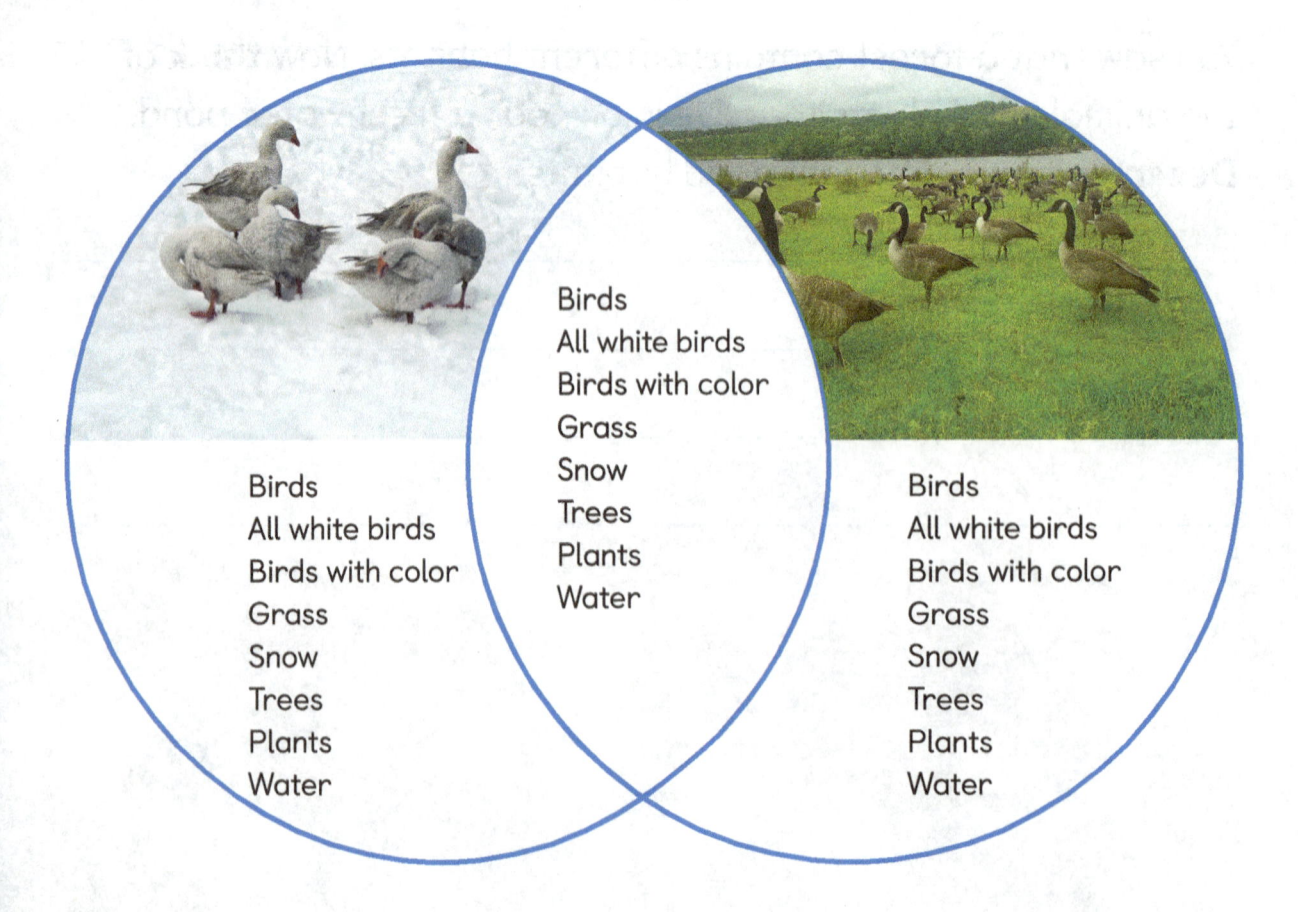

Left circle:
Birds
All white birds
Birds with color
Grass
Snow
Trees
Plants
Water

Overlap:
Birds
All white birds
Birds with color
Grass
Snow
Trees
Plants
Water

Right circle:
Birds
All white birds
Birds with color
Grass
Snow
Trees
Plants
Water

CCC Patterns

4.1: Living Landscapes

4.1 | Share   How do plants and animals meet their needs in a habitat?

Activity 26
Record Evidence Like a Scientist

Quick Code: us2784s

Climbing Perch

Now that you have learned about living landscapes, look again at the picture of the Climbing Perch. You first saw this in Wonder.

Let's Investigate Climbing Perch

Talk Together

How can you describe the climbing perch now? How is your explanation different from before?

SEP Constructing Explanations and Designing Solutions

66

Look at the Can You Explain? question. You first read this question at the beginning of the lesson.

> ### Can You Explain?
>
> How do plants and animals meet their needs in a habitat?

Now, you will use your new ideas about the Climbing Perch to answer a question.

1. **Choose** a question. You can use the Can You Explain? question, or one of your own. You can also use one of the questions that you wrote at the beginning of the lesson.

   Your Question

2. Then use the sentence starters to answer the question.

4.1: Living Landscapes | 67

4.1 | Share How do plants and animals meet their needs in a habitat?

Plants need
___
___
___

Plants meet their needs in a habitat by
___
___
___

I observed evidence of this when
___
___
___

Animals need

_____
_____
_____
_____

Animals meet their needs in a habitat by

_____
_____
_____
_____

I observed evidence of this when

_____
_____
_____
_____

STEM in Action

Quick Code: us2785s

Activity 27
Analyze Like a Scientist

Keeping Things Protected

Read the passage. **Watch** the video. Pay attention to the different parts of an ecosystem. **Look** at the image. The bison are in Yellowstone National Park. Use what you have read and seen to answer the questions.

Read Together

Keeping Things Protected

Ecosystems

Video

Smoke drifts out of a forest. Birds fly away. A forest fire started, but it was put out by the fire department. Now a park ranger has to help **restore** the area that was burned away by the forest fire. Imagine you are a park ranger and a forest fire has burned up a section of the park. What will you do?

SEP Obtaining, Evaluating, and Communicating Information **CCC** Cause and Effect

Park rangers are important to parks. Park rangers help make sure certain species can stay alive. After a forest fire, a park ranger clears away burned branches, grass, and leaves. A park ranger also plants new trees. This helps the forest grow back. New trees bring birds and other living things back to the park.

Bison in Yellowstone National Park

4.1: Living Landscapes

Have you visited a national park? What did you see there?

_____
_____
_____

How can national parks protect habitats?

_____
_____
_____

How can we take care of national parks?

_____
_____
_____

Maintaining the Park

Imagine that you are a park ranger that is working to maintain a park. You want to show some visitors to the park how some living and nonliving things are important to each other.

Look at the pictures of living and nonliving things. Each object pictured depends on one of the other objects. **Number** the pictures to show how each object depends on the next.

fish	
sun	
bear	
algae	

4.1: Living Landscapes

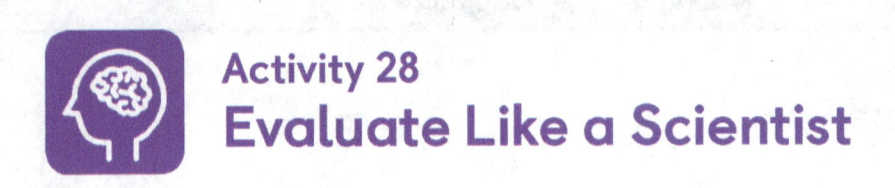

4.1 | Share
How do plants and animals meet their needs in a habitat?

Activity 28
Evaluate Like a Scientist

Quick Code: us2786s

Review: Living Landscapes

Think about what you have read and seen. What did you learn?

Draw what you have learned. Then, **tell** someone else about what you learned.

Talk Together

Think about what you saw in Get Started. Use your new ideas to talk about how to make a sports area that will protect plants and animals.

4.1: Living Landscapes | 75

CONCEPT
4.2

Plant and Animal Relationships

Student Objectives

By the end of this lesson:

☐ I can find examples of how plants and animals need each other. I can tell others about those examples.

☐ I can find out how the living things in a place affect each other. I can draw a diagram to show what I find out.

Key Vocabulary

☐ environment
☐ pollen
☐ pollination
☐ recycle
☐ shelter

Quick Code: us2788s

4.2: Plant and Animal Relationships

Activity 1
Can You Explain?

How do plants and animals work together?

_____

_____

_____

_____

Quick Code:
us2789s

4.2: Plant and Animal Relationships

4.2 | Wonder How do plants and animals work together?

Activity 2
Ask Questions Like a Scientist

Quick Code: us2790s

Mystery Image

Look at the photo. **Use** the photo to **play** a game. **Cover up** the photo. **Show** a partner one section of the photo at a time. Can your partner guess what the photo is showing?

Let's Investigate a Mystery Image

What are the brown, spiky objects?

_____

What is the black part of the image?

_____

How are the objects in the photo related?

_____

_____

What questions do you have about the mystery image?

Your Questions

4.2: Plant and Animal Relationships | 81

4.2 | Wonder — How do plants and animals work together?

Activity 3
Observe Like a Scientist

Quick Code: us2791s

Hiking Across a Woody Landscape

Look at the image. **Circle** the animal that was shown in the mystery image.

Hiking Across a Woody Landscape

Talk Together

Talk together about any other clues from the picture that may answer some of your questions.

Activity 4
Analyze Like a Scientist

Quick Code: us2792s

Impacting Landscapes

Read the passage and then, **answer** the question.

> Read Together

Impacting Landscapes

The mystery image shows a dog's black coat with a burr stuck to it. But there is more to the story.

A burr is a seed. It has spikes that stick to things. The dog was walking through the woods on a hike. When the dog walked by, the burr stuck to the dog's coat. The dog carried the burr seed to a new location.

4.2: Plant and Animal Relationships | 83

Read Together

A plant has roots that help it stay in one place. It also has seeds it needs to spread. The dog helps the plant spread seeds by carrying the burr to a new location.

The burrs will fall off the animal, or a human will remove the burrs from the dog. The burr seeds are now moved to a new location. The seeds can grow a new plant in the new location.

Plant with Burrs

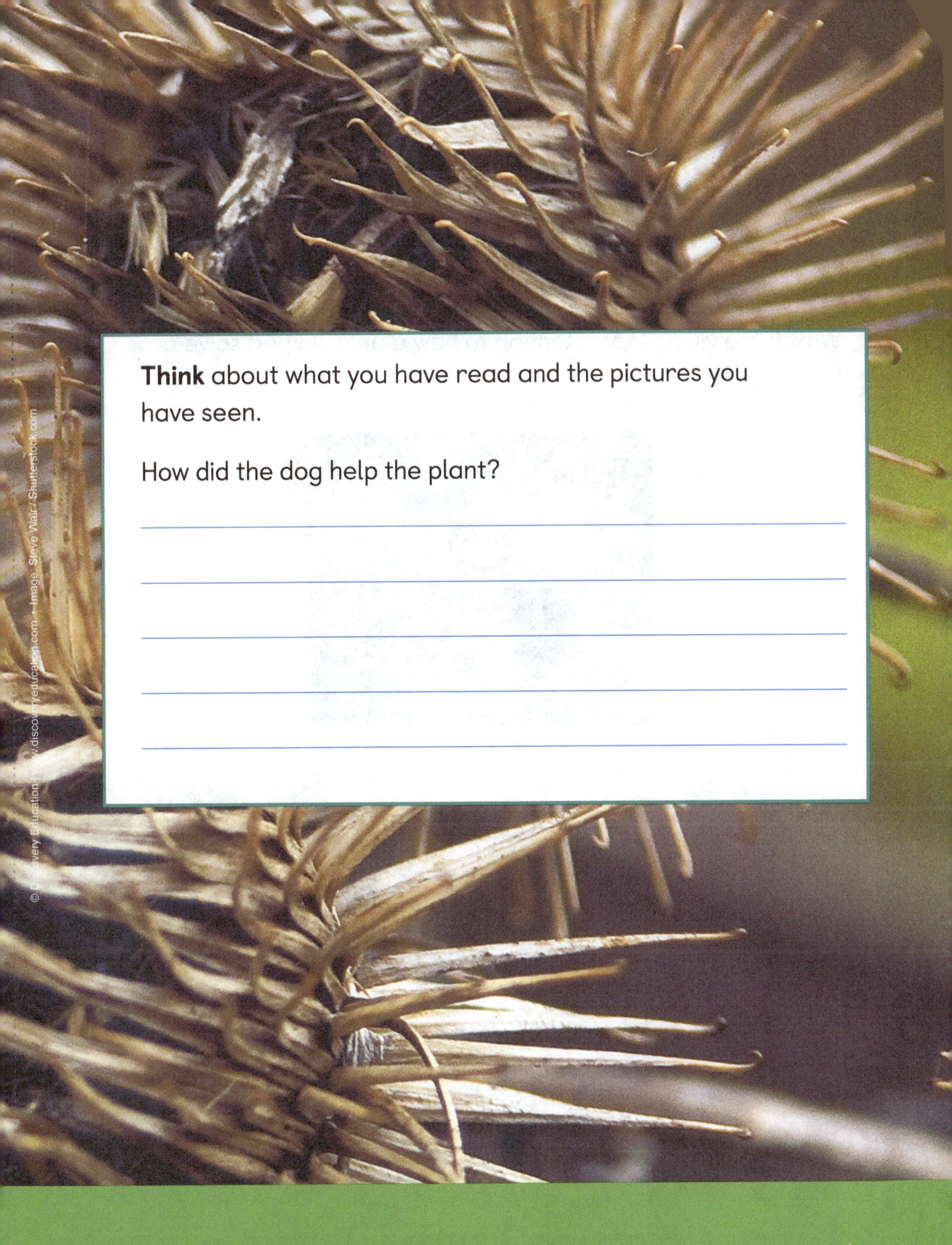

Think about what you have read and the pictures you have seen.

How did the dog help the plant?

_____

_____

_____

_____

_____

4.2 | Wonder How do plants and animals work together?

Activity 5
Observe Like a Scientist

Quick Code: us2793s

Hook and Loop

Watch the video. Pay attention to how a burr helped solve a human problem.

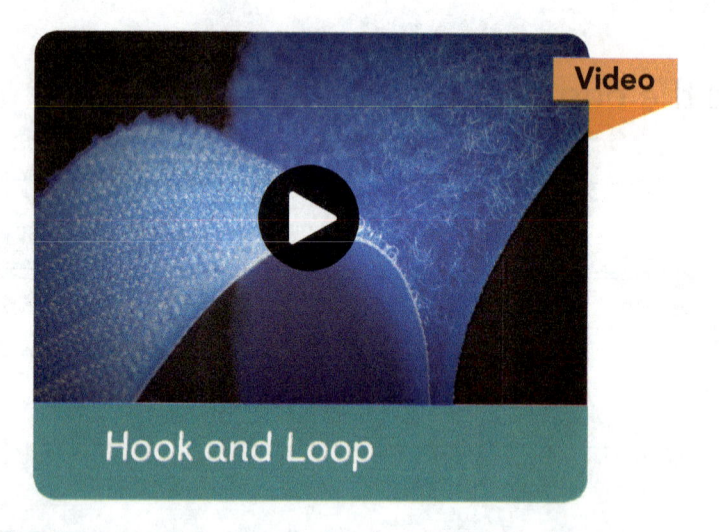

Video

Hook and Loop

Talk Together

Now, talk together about how the burr helped Mestral.

86

Activity 6
Observe Like a Scientist

Quick Code: us2794s

Bumblebee on Flower

Look at the image. **Think** about what you know about flowers and bees. Then, **answer** the questions.

Bumblebee on Flower

4.2: Plant and Animal Relationships | 87

4.2 | Wonder — How do plants and animals work together?

What do you know about flowers and bees?

_____

_____

_____

How does the bee meet the needs of the flower?

_____

_____

_____

How does the flower meet the needs of the bee?

_____

_____

_____

Activity 7
Evaluate Like a Scientist

Quick Code: us2795s

What Do You Already Know About Plant and Animal Relationships?

How Does an Apple Tree Grow?

Look at the pictures below. Each picture shows part of the cycle of apple tree growth. **Number** the pictures to show the order of growth. The seed has already been labeled number 1.

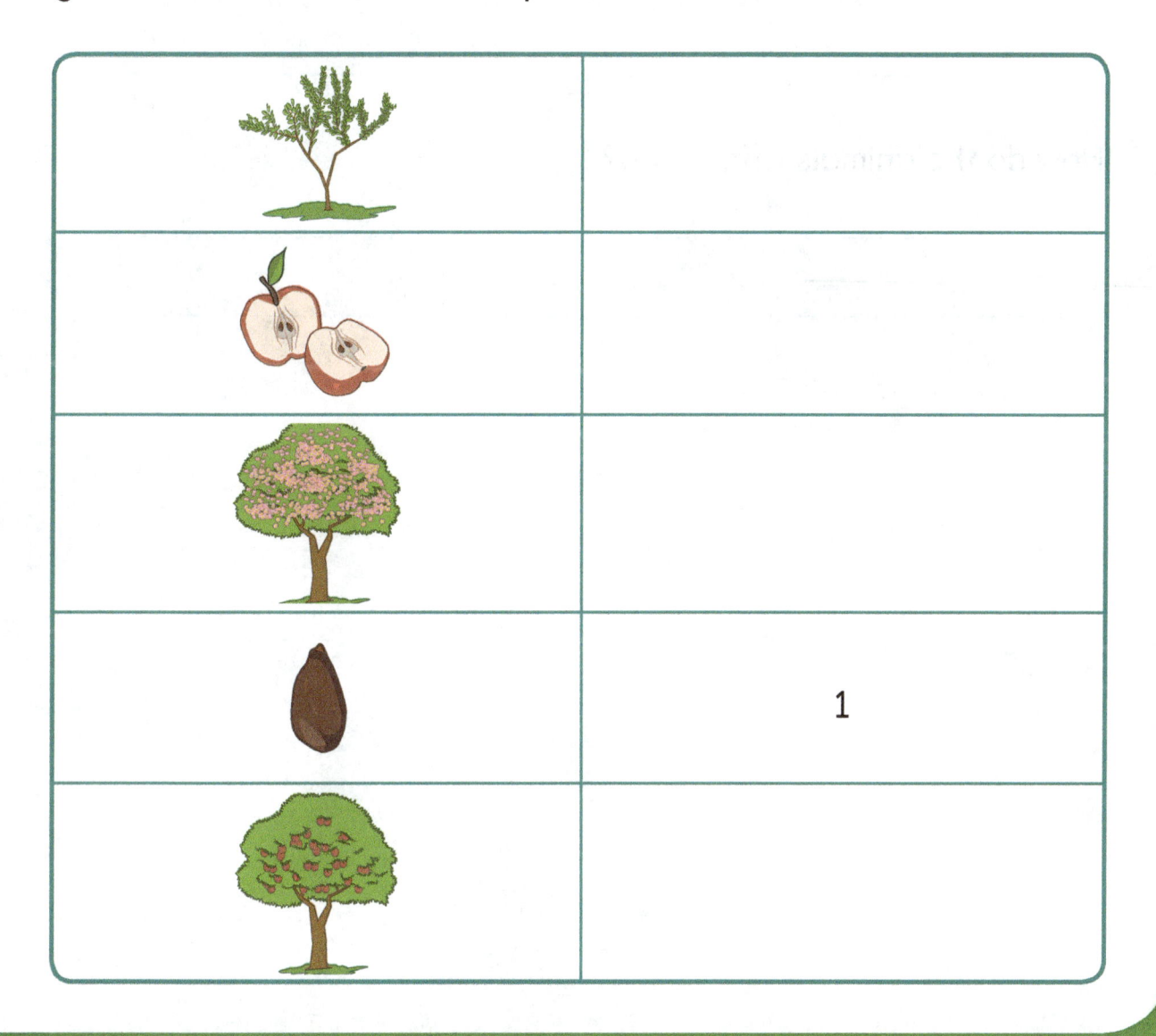

(young tree)	
(apple cut open)	
(tree with blossoms)	
(seed)	1
(tree with apples)	

4.2: Plant and Animal Relationships | 89

4.2 | Wonder   How do plants and animals work together?

Discussing Animals

Answer the questions to find out what you know about animals and people.

What animals live near you? List five animals.

_____

_____

How do you affect these animals?

_____

_____

How do the animals affect you?

_____

_____

4.2: Plant and Animal Relationships

Learn | How do plants and animals work together?

How Do Plants and Animals Rely on Each Other?

Activity 8
Observe Like a Scientist

Quick Code: us2796s

The Bee

Watch the video. Pay attention to how bees help plants.

Write your thoughts about what you observed in the video.

The Fixies: The Bee

Use the AEIOU strategy to show your thinking.

A = Adjective (describing word)

_____

E = Emotion (feeling word)

_____

I = Interesting (something of interest)

_____

O = Oh! (something surprising)

_____

U = Um? (a question)

_____

Talk Together

Now, talk together about the experiences you have had with bees.

4.2: Plant and Animal Relationships

Read Together

How Do Plants and Animals Depend on Each Other?

Have you ever been scared by a bee? Bees may be scary to some people, but they are very helpful insects. Many plants could not survive without bees.

Bees help plants by taking **pollen** from one plant to another plant. This is called **pollination**. Pollination helps plants reproduce to make new plants.

Bees are not the only animals that help with pollination. Hummingbirds, butterflies, and bats also help with this task.

Bumblebee on Flower

After pollination, a plant begins to grow a fruit. Inside the fruit is a seed. Once again, animals help plants. How do they do this?

Activity 9
Analyze Like a Scientist

Quick Code: us2797s

How Do Plants and Animals Depend on Each Other?

Think about what you have read and the pictures you have seen.

Besides bees, what are other ways pollen can travel between plants?

_____
_____
_____

4.2: Plant and Animal Relationships | 95

4.2 | Learn How do plants and animals work together?

Activity 10
Think Like a Scientist

Quick Code: us2798s

Pollinators

In this activity, you will create a model of an animal that pollinates plants. Then, you will test your model.

What materials do you need? (per group)

- Paper
- Pencils
- Modeling clay
- Scissors
- Access to Internet (optional)
- Images of various flowers and pollinators (print, online, or from DE Streaming; for example, bat, bee, hummingbird, butterfly, lemur, or mouse)
- Chenille sticks
- Cotton swabs
- Plastic container, 12 oz
- Masking tape
- Tissue paper
- Glue

What Will You Do?

1. **Look** at images of various pollinators and flowers.

2. **Research** your pollinator.

SEP Developing and Using Models

3. **Draw** and **label** a sketch of your model.

Your Pollinator

4.2: Plant and Animal Relationships | 97

4.2 | Learn How do plants and animals work together?

4. **Build** your models.

5. **Plan** a test for the model and **carry** out the test.

6. **Record** the results of your test.

How is your model similar to an actual pollinator and plant?

_____
_____
_____
_____

How is your model different from an actual pollinator and plant?

_____
_____
_____
_____

4.2 | Learn — How do plants and animals work together?

How is your model similar to other groups' models?

_____
_____
_____
_____

How is your model different from other groups' models?

_____
_____
_____
_____

How could you improve your model to make it more like a real pollinator and plant?

_____
_____
_____
_____

Activity 11
Observe Like a Scientist

Quick Code: us2799s

Building a Model of How Some Seeds Are Spread

Watch the video. **Look** at the shape of the seeds. Pay attention to how the seeds travel on the wind.

Video

Seed Dispersion: Building a Model of How Some Seeds Are Spread

Talk Together

Now, talk together about how you can model the way seeds are spread.

SEP Developing and Using Models

4.2: Plant and Animal Relationships

4.2 | Learn   How do plants and animals work together?

Activity 12
Observe Like a Scientist

Quick Code: us2800s

Types of Seeds

Look at the photos. **Look** at the shape of the seeds. **Think** about how the seeds will travel.

| Thistle Seed | Mangrove Seed | Maple Seed |

Talk Together

Now, talk together about how seed shapes and characteristics affect how seeds travel.

CCC   Structure and Function

102

Activity 13
Observe Like a Scientist

Quick Code: us2801s

Organism Needs

Do the interactive. **Build** a habitat for the animal. **Find** its land cover, food, and **shelter**. Then, **write** what you did in the data table.

Interactive

Organism Needs

SEP Planning and Carrying Out Investigations

4.2: Plant and Animal Relationships | 103

4.2 | Learn   How do plants and animals work together?

Animal	Land Cover	Food	Shelter
Lion			
Parrot			
Beaver			

Answer the questions.

What are three things that all animals need to live?

_____

Parrots often build their nests in tree hollows. How would a tree hollow protect a parrot and its chicks?

_____

_____

_____

How does the lion's habitat help it hunt?

_____

_____

_____

What do you think happens to animals if their habitat changes or is harmed in some way?

For example, what would happen to beavers if people drained marshland or cut down a forest to build new houses?

_____

_____

_____

_____

4.2: Plant and Animal Relationships

4.2 | Learn — How do plants and animals work together?

Activity 14
Investigate Like a Scientist

Quick Code: us2802s

Hands-On Investigation: Traveling Seeds

In this activity, you will design and test models of how seeds move from place to place.

Make a Prediction

You are going to test different ways seeds can move from one place to another place. **Write** or **draw** your predictions.

Which type of seed do you think will be the easiest to move from one place to another?

SEP Developing and Using Models   **CCC** Structure and Function

What materials do you need? (per group)

- Paper
- Pencils
- Modeling clay
- Felt
- Artificial feathers
- Fake fur
- Tissue paper
- Toothpicks
- Sequins
- Chenille sticks
- Masking tape
- Cotton balls

What Will You Do?

Choose three seeds you are going to model.

Draw and **label** a sketch of the seeds.

Label your seed drawings with the materials you will use to build your model.

Your Pollinator

4.2: Plant and Animal Relationships

4.2 | Learn   How do plants and animals work together?

Build your models.

Plan a test for the model and **carry** out the test.

Record the results of your test.

Think About the Activity

Describe how the parts of the seeds you modeled helped them move from one place to another place.

_____
_____
_____
_____

What kinds of seeds do you think are the easiest to move from place to place? Why?

_____
_____
_____
_____

How could you have improved the testing of your seed models?

_____
_____
_____
_____

4.2: Plant and Animal Relationships

4.2 | Learn — How do plants and animals work together?

Activity 15
Observe Like a Scientist

Quick Code: us2803s

Insects Are Helpful

Watch the video. Pay attention to the pollination of plants by insects.

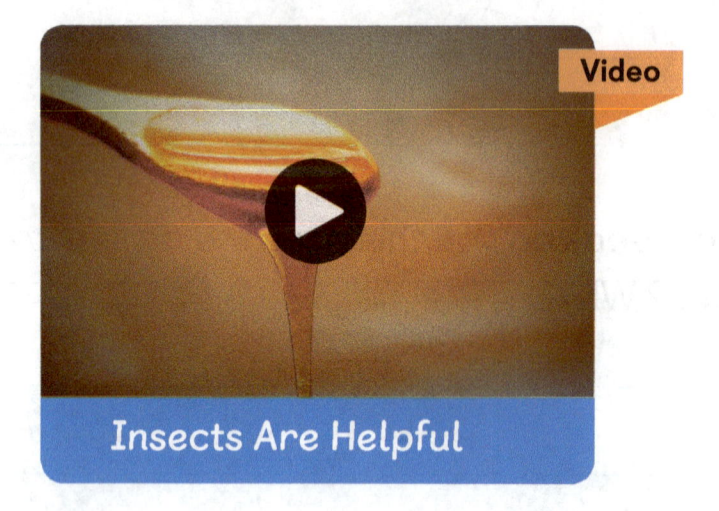

Insects Are Helpful (Video)

Talk Together

Now, talk together about how insects pollinate and why it is important.

CCC Cause and Effect

Draw and **write** what you learned from the video.

Drawing	Writing

4.2: Plant and Animal Relationships

Activity 16
Analyze Like a Scientist

Quick Code: us2804s

How Plants and Animals Help Each Other

Read the passage and **look** at the pictures. Then **underline** any sentences with examples of plants and animals helping each other.

> Read Together

How Plants and Animals Help Each Other

Animals help plants with pollination by carrying seeds from place to place. Animals like squirrels spread seeds when they hide them to eat later. Sometimes seeds stick to an animal's coat. Other times, an animal eats the fruit with seeds. The seeds exit the animal's body in new locations.

Squirrel Eating a Nut

CCC Cause and Effect

How do plants help animals? Some animals eat plants for food. Animals also use plants for shelter. A beaver uses logs from trees to build a lodge. A bird uses sticks to build a nest.

Plants need animals. Animals need plants. Both plants and animals help each other live and survive.

Beaver Lodge

4.2 | Learn How do plants and animals work together?

Activity 17
Evaluate Like a Scientist

Quick Code: us2805s

Helping Out

Animals and plants help each other in their habitats. **Read** the following sentences. **Circle** the sentences that tell how animals and plants help each other in their habitats.

- Humans eat vegetables, and they water vegetables to help them grow.

- A bear eats a fish.

- A plant called a Venus flytrap eats a grasshopper.

- Bees get nectar from flowers and pollinate other flowers as they gather more nectar.

CCC Cause and Effect

How Do Humans Change the Habitats Where Plants and Animals Live?

Activity 18
Observe Like a Scientist

Quick Code: us2806s

Litter on the Beach

Look at the image. **Think** about where the trash came from. Then, **answer** the questions.

Litter on the Beach

4.2: Plant and Animal Relationships | 115

4.2 | Learn — How do plants and animals work together?

Would you like to visit this beach? Why or why not?

_____
_____
_____
_____

How does trash end up in the water?

_____
_____
_____
_____

In what other places have you seen trash?

_____
_____
_____

Activity 19
Analyze Like a Scientist

Quick Code: us2807s

How Do Humans Change the Habitats Where Plants and Animals Live?

Read the passage about what happens when animals lose most of their habitats. Then, **complete** the activity that follows.

Read Together

How Do Humans Change the Habitats Where Plants and Animals Live?

Humans change their **environment** in many ways. What happens when humans make changes to the environment? Sometimes, plants and animals can adjust to the changes.

Humans make large and small changes to habitats. Plants and animals can adjust to some changes to their habitats. If a human makes a large change to the habitat, the plants and animals may not be able to adjust.

CCC Cause and Effect

4.2: Plant and Animal Relationships | 117

Read Together

When humans leave trash on the ground, it is called litter. Litter is a form of pollution. Pollution destroys habitats.

Plants and animals need food, water, and shelter. Changes to habitat environments may destroy these resources. This can cause problems for the plants and animals that live there.

A Forest Destroyed

Human activities can change a habitat. Humans cut down trees to make room for farmland and buildings. This causes plants and animals to lose their homes. People use chemicals for farming and factories. These can pollute the land, air, and water. Chemicals can cause plants and animals to die. When campfires are not put out properly, they can cause forest fires that can destroy a habitat.

4.2: Plant and Animal Relationships

Think about how people can harm a habitat. **Think** about how people can help a habitat. **Write** your ideas in the chart.

Harm	Help

Activity 20
Observe Like a Scientist

Quick Code: us2808s

Rain Forests in Danger

Watch the video. **Look** for plants and animals in the rain forests. Pay attention to how people are changing rain forests. Then, **talk** about the question.

Video

Rain Forests in Danger

Talk Together

Now, talk together about how humans are changing the rain forests and affecting the animals and plants. What are the effects of these changes?

CCC Cause and Effect

4.2: Plant and Animal Relationships | 121

Activity 21
Analyze Like a Scientist

Quick Code: us2809s

Animals Without a Home

Read the passage about what happens when animals lose most of their habitats.

Then, **complete** the activity that follows.

> Read Together

Animals Without a Home

Those trees are homes for animals like birds. They have nowhere else to live. They need the trees to live. People cut down trees to build shops and homes. The animals lose their homes.

Owls Perching in a Tree

SEP Obtaining, Evaluating, and Communicating Information

Imagine your home was replaced with a parking lot. You have no place to live. Not having a home happens to many animals each day.

Some animals have lost most of their habitats. We can help them stay alive. We can protect them and their homes. We can make new homes for them.

Bobcat Resting in a Tree

4.2: Plant and Animal Relationships

Pretend you are an animal that recently lost its home because trees were cut down to make a parking lot. **Think** about a story you could write about the animal. What would the animal do?

Write notes for your story in the story map.

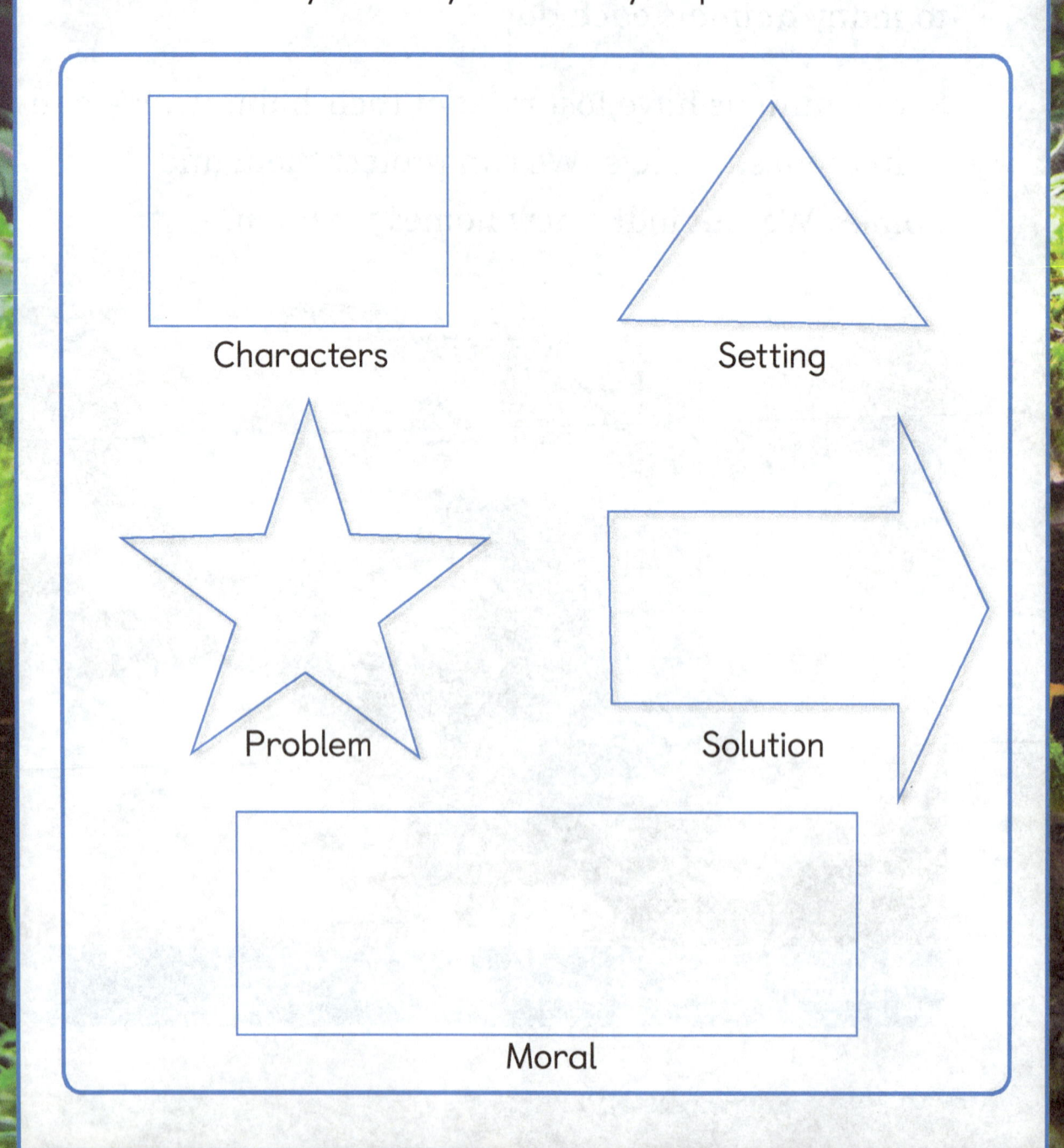

Characters

Setting

Problem

Solution

Moral

Write a story about the animal. **Say** what the animal would do.

4.2 | Learn How do plants and animals work together?

Activity 22
Observe Like a Scientist

Quick Code: us2810s

Polluted Run-Off Song

Watch the video. **Listen** to find out ways that people can harm the environment.

Polluted Run-Off Song

Talk Together

Now, talk together about how some things you do every day can harm the environment. Talk together about how you could do these things less often or in a different way.

SEP Constructing Explanations and Designing Solutions

SEP Obtaining, Evaluating, and Communicating Information

What would you do to help the environment? **Draw** or **write** your ideas.

4.2: Plant and Animal Relationships

4.2 | Learn   How do plants and animals work together?

Make a poster to show others ways they can help the environment.

Your Poster

Activity 23
Observe Like a Scientist

Quick Code: us2811s

Keeping Our Environment Clean

Watch the video. **Listen** to find out how human activities affect the environment. Then, **talk** about the questions.

Keeping Our Environment Clean

Talk Together

Now, talk together about two questions.

- How do humans affect the environment?
- How does the environment affect humans?

4.2: Plant and Animal Relationships | 129

4.2 | Learn — How do plants and animals work together?

Activity 24
Evaluate Like a Scientist

Quick Code: us2812s

Habitat Problems

Look at the pictures. Pay attention to how the habitat is being hurt.

Image 1

Image 2

Image 3

CCC Cause and Effect

Write the number of the correct image next to the animal it will affect.

4.2 | Share How do plants and animals work together?

Activity 25
Record Evidence Like a Scientist

Quick Code: us2813s

Mystery Image

Now that you have learned about plant and animal relationships, look again at the picture of the Mystery Image. You first saw this in Wonder.

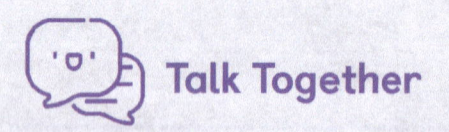

Let's Investigate a Mystery Image

Talk Together

How can you describe the Mystery Image now? How is your explanation different from before?

SEP Constructing Explanations and Designing Solutions

Look at the Can You Explain? question. You first read this question at the beginning of the lesson.

> ### Can You Explain?
> How do plants and animals work together?

Now, you will use your new ideas about Let's Investigate a Mystery Image to answer a question.

1. **Choose** a question. You can use the Can You Explain? question, or one of your own. You can also use one of the questions that you wrote at the beginning of the lesson.

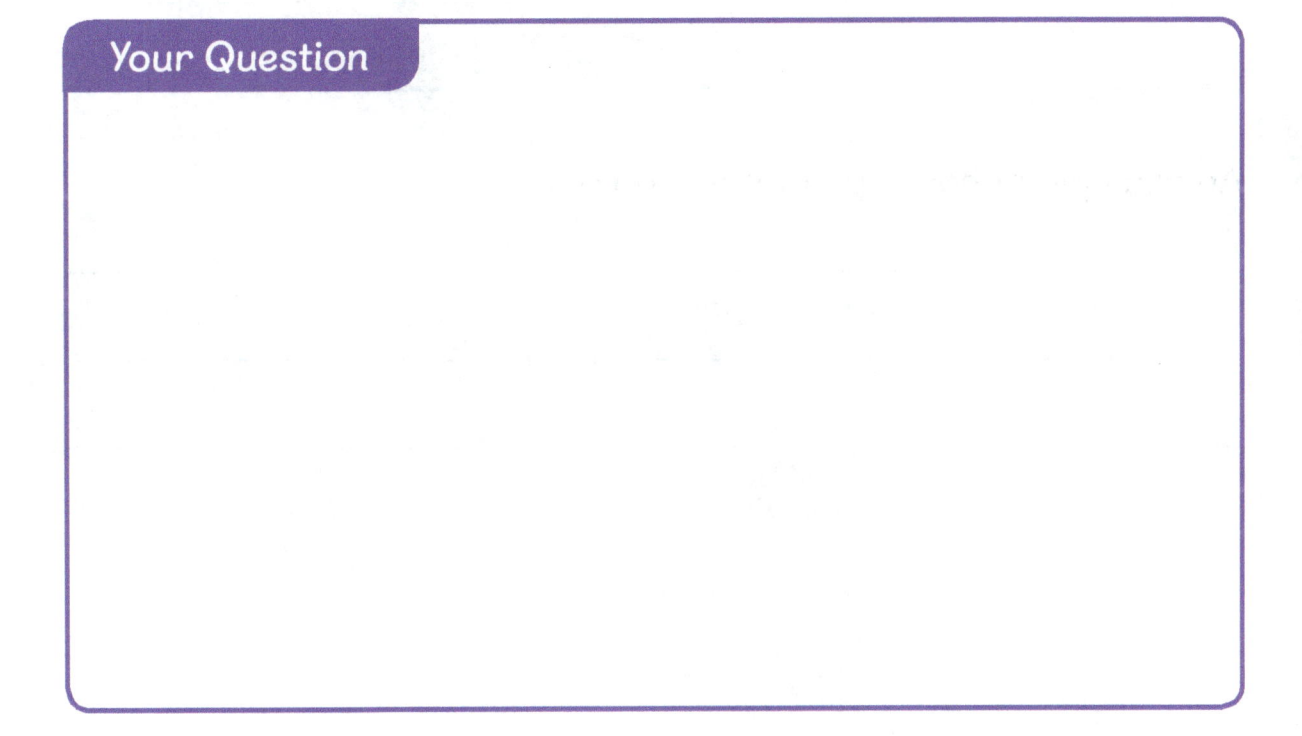

   Your Question

2. Then use the sentence starters to answer the question.

4.2: Plant and Animal Relationships | 133

4.2 | Share — How do plants and animals work together?

The burr sticks to the dog's fur because

_____

_____

Evidence that would support my claim for the burr is

_____

_____

An example of how an animal helps a plant is

_____

_____

An example of how a plant helps an animal is

_____

_____

_____

Some ways that plants and animals depend on each other are

_____

_____

_____

_____

Some ways humans can affect the habitats where plants and animals live are

_____

_____

_____

_____

4.2: Plant and Animal Relationships | 135

STEM in Action

Quick Code: us2814s

Activity 26
Analyze Like a Scientist

Making a Home for Animals

Read about making a home for animals. Then, **answer** the questions.

> **Read Together**

Making a Home for Animals

Have you ever gone to a new park? Maybe you saw a squirrel with a nut. Maybe you saw a bird building a nest. These animals are having their needs met.

What might happen if they were not being met? Perhaps people cut down all the trees in their home. Maybe a fire burned down their meadow. Sometimes, these animals can be saved. They are taken to a zoo. While at the zoo, the zookeepers make sure their needs are met. How does a zookeeper plan an animal's home?

SEP Obtaining, Evaluating, and Communicating Information

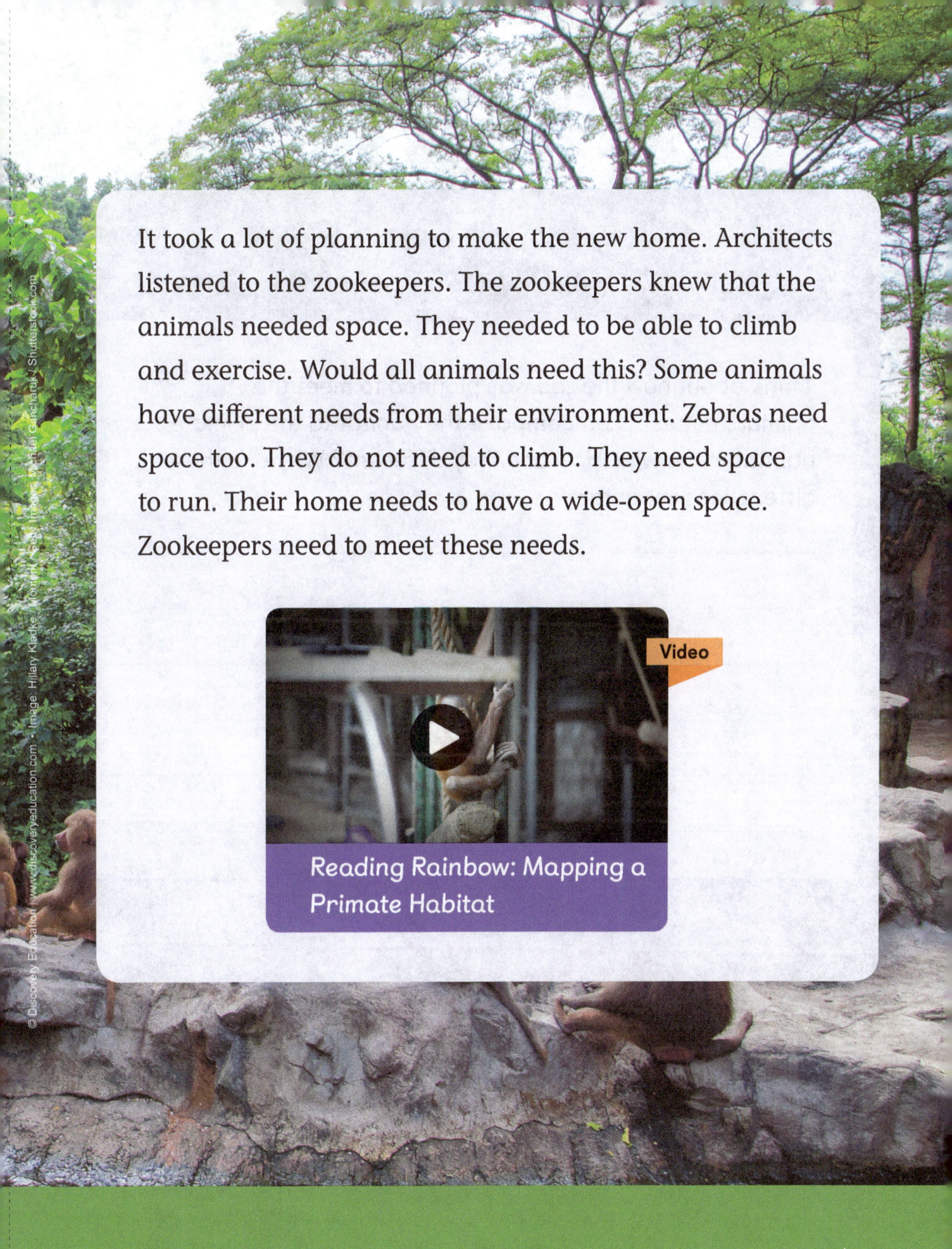

It took a lot of planning to make the new home. Architects listened to the zookeepers. The zookeepers knew that the animals needed space. They needed to be able to climb and exercise. Would all animals need this? Some animals have different needs from their environment. Zebras need space too. They do not need to climb. They need space to run. Their home needs to have a wide-open space. Zookeepers need to meet these needs.

Reading Rainbow: Mapping a Primate Habitat

4.2: Plant and Animal Relationships

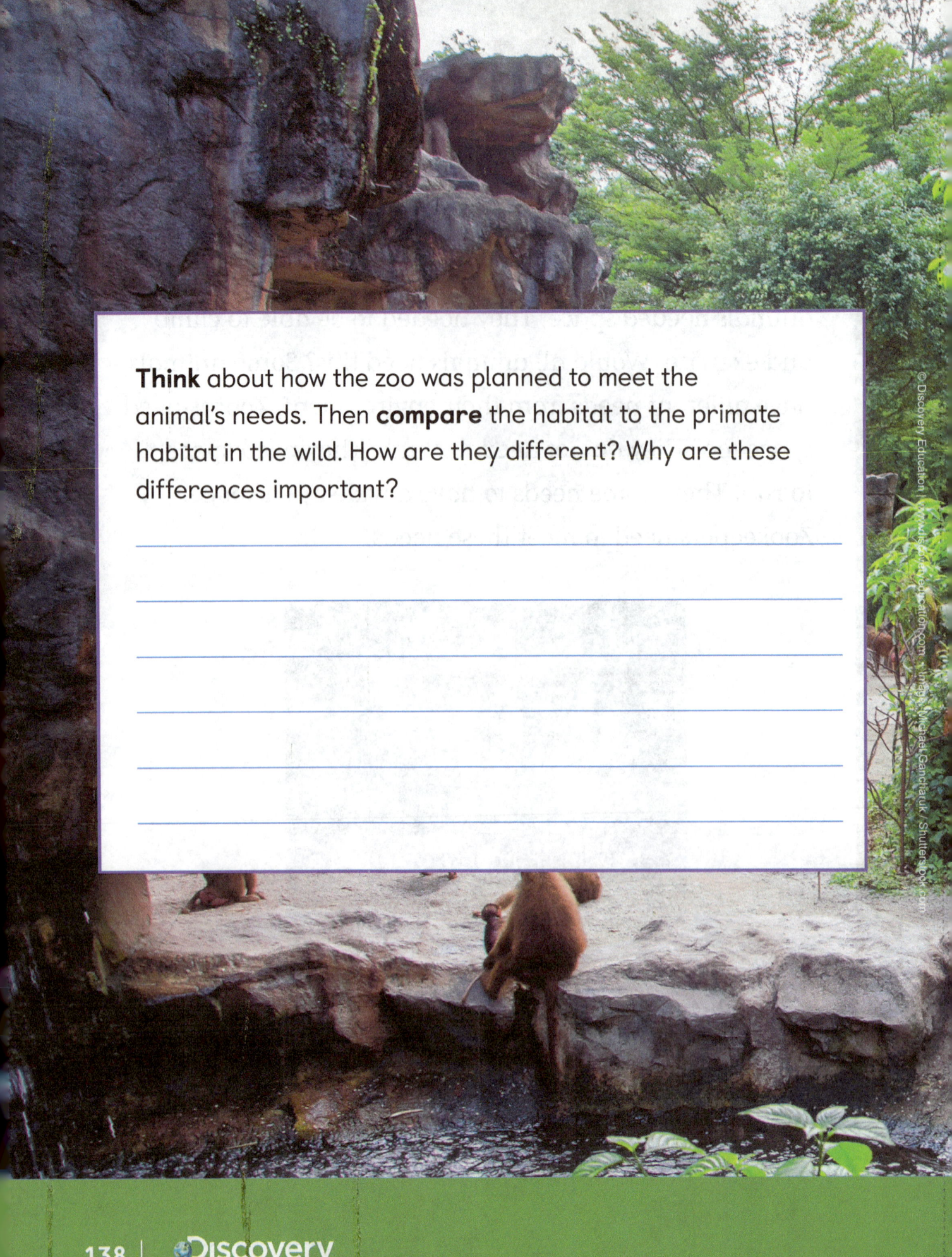

Think about how the zoo was planned to meet the animal's needs. Then **compare** the habitat to the primate habitat in the wild. How are they different? Why are these differences important?

_____

_____

_____

_____

_____

_____

New Pet

Drew bought a lizard from the pet store. He needs to put it in a terrarium. He wants to make sure he buys the correct supplies for the lizard's habitat. Drew only has $5.00 total to spend on the habitat for the lizard. The lizard needs both plants and a lamp.

Habitat Supply	Cost
Lamp	$2.50
Plant	$1.25
Sand	$6.00

How many plants and lamps can Drew buy with his $5.00?

_____

_____

_____

_____

_____

4.2: Plant and Animal Relationships

4.2 | Share How do plants and animals work together?

Activity 27
Evaluate Like a Scientist

Quick Code: us2815s

Review: Plant and Animal Relationships

Think about what you have read and seen. What did you learn? **Draw** what you have learned. Then, **tell** someone else about what you learned.

Talk Together

Think about what you saw in Get Started. Use your new ideas to talk about how to make a sports area that will protect plants and animals.

4.2: Plant and Animal Relationships | 141

Unit Project

Design Solutions Like a Scientist

Quick Code: us2816s

Hands-On Engineering: Eco-Friendly Outdoor Area

In this activity, you will design an eco-friendly outdoor sports area. To complete the activity, you will use what you know about how living things depend on one another to survive.

Playing Soccer on a Soccer Field

SEP	Asking Questions and Defining Problems	**SEP**	Constructing Explanations and Designing Solutions
SEP	Developing and Using Models	**CCC**	Cause and Effect
SEP	Planning and Carrying Out Investigations	**CCC**	Systems and System Models

What materials do you need? (per group)
- An outdoor area that students can alter and change
- A variety of materials to design an outdoor space (based on students' materials lists)

HANDS-ON ENGINEERING

Ask Questions About the Problem

Look at the outdoor area.

What living things might depend on the plants in the area?

_____
_____
_____

How can your design prevent harm to the living things in the area?

_____
_____
_____

Unit Project

What Will You Do?

Brainstorm some ideas for your outdoor sports area.
Draw a few of your ideas.

What surface do you need to play your sport? What changes will you need to make to the area to play your sport?

How will you know if your design protects living things in the sports area?

_____
_____
_____
_____

Test your design. **Draw** a picture or use words to show how you tested your design.

Unit Project

Think About the Activity

Write or **draw** your answers to the questions in the chart.
How well did your design reduce the effect on the environment?
How could you improve your design?

What Worked?	What Didn't Work?

What Could Work Better?

Grade 2 Resources

- Bubble Map
- Safety in the Science Classroom
- Vocabulary Flash Cards
- Glossary
- Index

Name _____

Bubble Map

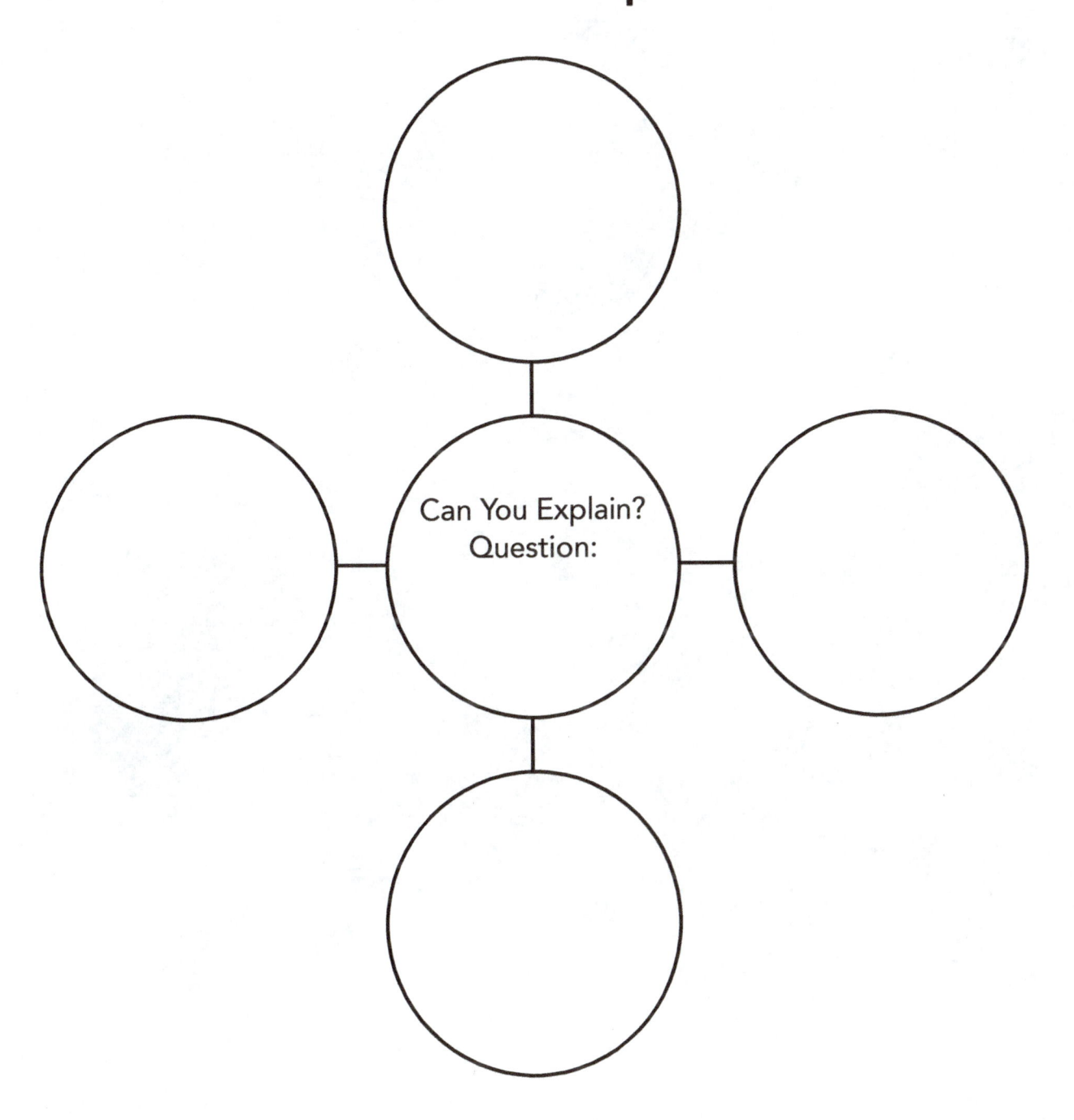

Can You Explain?
Question:

Bubble Map | R1

Safety in the Science Classroom

Following common safety practices is the first rule of any laboratory or field scientific investigation.

Dress for Safety

One of the most important steps in a safe investigation is dressing appropriately.

- Splash goggles need to be kept on during the entire investigation.

- Use gloves to protect your hands when handling chemicals or organisms.

- Tie back long hair to prevent it from coming in contact with chemicals or a heat source.

- Wear proper clothing and clothing protection. Roll up long sleeves, and if they are available, wear a lab coat or apron over your clothes. Always wear closed-toe shoes. During field investigations, wear long pants and long sleeves.

Safety Goggles

Be Prepared for Accidents

Even if you are practicing safe behavior during an investigation, accidents can happen. Learn the emergency equipment location in your classroom and how to use it.

- The eye and face wash station can help if a harmful substance or foreign object gets into your eyes or onto your face.

- Fire blankets and fire extinguishers can be used to smother and put out fires in the laboratory. Talk to your teacher about fire safety in the lab. He or she may not want you to directly handle the fire blanket and fire extinguisher. However, you should still know where these items are in case the teacher asks you to retrieve them.

Most importantly, when an accident occurs, immediately alert your teacher and classmates. Do not try to keep the accident a secret or respond to it by yourself. Your teacher and classmates can help you.

Fire Extinguisher

Practice Safe Behavior

There are many ways to stay safe during a scientific investigation. You should always use safe and appropriate behavior before, during, and after your investigation.

- Read all of the steps of the procedure before beginning your investigation. Make sure you understand all the steps. Ask your teacher for help if you do not understand any part of the procedure.

- Gather all your materials and keep your workstation neat and organized. Label any chemicals you are using.

- During the investigation, be sure to follow the steps of the procedure exactly. Use only directions and materials that have been approved by your teacher.

- Eating and drinking are not allowed during an investigation. If asked to observe the odor of a substance, do so using the correct procedure known as wafting, in which you cup your hand over the container holding the substance and gently wave enough air toward your face to make sense of the smell.

- When performing investigations, stay focused on the steps of the procedure and your behavior during the investigation. During investigations, there are many materials and equipment that can cause injuries.

- Treat animals and plants with respect during an investigation.

- After the investigation is over, appropriately dispose of any chemicals or other materials that you have used. Ask your teacher if you are unsure of how to dispose of anything.

- Make sure that you have returned any extra materials and pieces of equipment to the correct storage space.

- Leave your workstation clean and neat. Wash your hands thoroughly.

Vocabulary Flash Cards

biodiversity
the many different types of life that live together in an environment

environment
all the living and nonliving things that surround an organism

habitat
the place where a plant or animal lives

landscape
the view of a land's surface

Vocabulary Flash Cards | R7

nutrient

something in food that helps people, animals, and plants live and grow

organism

a living thing

pollen

the yellow powder found inside a flower

pollination

moving or carrying pollen from a plant to make the seeds grow

Vocabulary Flash Cards | R9

recycle

to create new materials from something already used

restore

to put into use again

shelter

a place that protects you from harm or bad weather

Vocabulary Flash Cards | R11

Glossary

English ———— A ———— Español

absorb
to take in or soak up

absorber
tomar o captar

absorption
how much something can take in and hold

absorción
cuanto algo puede tomar y retener

adjust
to fix or change something

ajustar
arreglar o cambiar algo

analyze
to closely examine something and then explain it

analizar
examinar con atención algo y luego explicarlo

———— B ————

barrier
something that is used to stop or block materials from moving

barrera
algo que se usa para evitar o bloquear el movimiento de materiales

biodiversity
the many different types of life that live together in an environment

biodiversidad
muchos y diferentes tipos de vida que conviven en un medio ambiente

--- C ---

canyon
a deep valley that has very steep sides

cañón
valle profundo que tiene laderas muy pronunciadas

channel
a path that is dug and used for drainage or protection against things like water, mud, or rocks

canal
vía cavada que se usa como desagüe o protección contra cosas como el agua, el lodo o las rocas

characteristic
a special quality that something may have

característica
cualidad especial que tiene algo

--- D ---

dissolve
to mix something with a liquid, such as water, so that it can't be seen anymore

disolver
mezclar algo con un líquido, como el agua, de manera que no se pueda ver más

drought
when there is no rain for a long period

sequía
cuando no llueve durante un período prolongado

---- E ----

Earth's crust
the top layer of Earth that is the thinnest and the most important because it is where we live

corteza de la Tierra
capa superior de la Tierra que es la más delgada y la más importante porque allí es donde vivimos

earthquake
a sudden shaking of the ground caused by the movement of rock underground

terremoto
repentina sacudida de la tierra causada por el movimiento de roca subterránea

elevation
the height of an area of land above sea level

elevación
altura de un área de tierra por encima del nivel del mar

engineer
a person who designs something that may be helpful to solve a problem

ingeniero
persona que diseña algo que puede ser útil para resolver un problema

engineering
using math and science to design and build machines, structures, and other devices

ingeniería
usar las matemáticas y las ciencias para diseñar y construir máquinas, estructuras y otros dispositivos

environment
all the living and nonliving things that surround an organism

medio ambiente
todos los seres vivos y objetos sin vida que rodean a un organismo

erosion
when soil is moved from one location to another by wind or water

erosión
cuando el viento o el agua transporta suelo de un lugar a otro

estimate
to make a careful guess

estimar
hacer una suposición consciente

---------- F ----------

feature
a thing that describes what something looks like; part of something

rasgo
cosa que describe cómo se ve algo; parte de algo

flexibility
the ability to bend without breaking

flexibilidad
capacidad de doblarse sin romperse

---

fresh water
water that is not salty, such as that found in streams and lakes

agua dulce
agua que no es salada, como la que se encuentra en arroyos y lagos

---

G

gemstone
a colorful stone found in nature that can be used for jewelry

piedra preciosa
piedra colorida que se encuentra en la naturaleza y se puede usar para hacer joyas

---

H

habitat
the place where a plant or animal lives

hábitat
lugar donde vive una planta o un animal

---

hardness
a measure of how difficult it is to scratch a mineral: Diamonds are the hardest mineral. They have a hardness scale rating of 10.

dureza
medida de cuán difícil es rayar un material: los diamantes son los minerales más duros. Su clasificación en la escala de dureza es 10.

Glossary | R17

L

landfill
a place where trash is buried

vertedero
lugar donde se entierra la basura

landform
a feature of Earth that has been formed by nature, such as a hill or a valley

accidente geográfico
característica de la Tierra formada por la naturaleza, como una colina o un valle

landscape
the view of a land's surface

paisaje
vista de la superficie de un terreno

location
a place where something is

ubicación
lugar donde se encuentra algo

M

map
a flat picture or drawing of a place that is made to show things, such as streets or towns, in an area

mapa
imagen o dibujo plano de un lugar que se hace para mostrar cosas, como las calles o las ciudades, de un área

material
things that can be used to build or create something

material
cosas que se pueden usar para construir o crear algo

matter
the things around you that take up space like solids, liquids, and gases

materia
cosas que nos rodean y ocupan espacio, como los sólidos, los líquidos y los gases

mixture
a combination of different things, but you can pick out each different one

mezcla
combinación de diferentes cosas, pero se puede identificar cada una

model
a human-made version created to show the parts of something else, either big or small

modelo
versión creada por el hombre para mostrar las partes de algo más, ya sea grande o pequeño

mountain
a very tall area of land that is higher than a hill and has steep sides

montaña
área de tierra muy alta que es más alta que una colina y tiene laderas pronunciadas

N

naturalist
someone who studies nature, especially plants and animals

naturalista
alguien que estudia la naturaleza, especialmente las plantas y los animales

nutrient
something in food that helps people, animals, and plants live and grow

nutriente
algo en los alimentos que ayuda a las personas, los animales y las plantas a vivir y crecer

O

observe
to watch closely

observar
mirar atentamente

ocean
a large body of salt water

océano
gran cuerpo de agua salada

organism
a living thing

organismo
ser vivo

P

plain
a large flat area of land without trees

llanura
gran área de tierra llana sin árboles

plateau
a large, flat area of land that is higher than the other land around it

meseta
gran área de tierra llana que está a más altura que el terreno que la rodea

pollen
the yellow powder found inside a flower

polen
polvo amarillo que se encuentra dentro de una flor

pollination
moving or carrying pollen from a plant to make the seeds grow

polinización
transferencia o transporte de polen de una planta para hacer que crezcan las semillas

preserve
to protect or keep something safe

preservar
proteger o mantener algo a salvo

property
a characteristic of something

propiedad
característica de algo

---- Q ----

quadrilateral
a flat shape with four straight sides, such as a square or a parallelogram

cuadrilátero
figura plana con cuatro lados rectos, como un cuadrado o un paralelogramo

R

recycle
to create new materials from something already used

reciclar
crear nuevos materiales a partir de algo usado

relief map
a type of map that shows how flat or steep the landforms are in an area

mapa de relieve
tipo de mapa que muestra si los accidentes geográficos son llanos o pronunciados en un área

resource
a material that can be used to solve problems

recurso
material que se puede usar para resolver problemas

restore
to put into use again

restablecer
volver a poner en servicio

reverse engineering
the process of learning about something by taking it apart to see how it works and what it is made of

ingeniería inversa
proceso de aprender acerca de algo, desarmándolo para ver cómo funciona y de qué está hecho

river
water flowing through a landscape, usually fed by smaller streams

río
agua que fluye a través de un área, por lo general alimentada por arroyos más pequeños

S

select
to choose or pick

seleccionar
elegir o escoger

shelter
a place that protects you from harm or bad weather

refugio
lugar para protegerse de peligros o el mal tiempo

slope
land that is slanted or angles downward

pendiente
tierra inclinada hacia abajo

soil
dirt that covers Earth, in which plants can grow and insects can live

suelo
tierra que cubre nuestro planeta en la que pueden crecer plantas y vivir insectos

Glossary | R23

solution
a combination of two things that are mixed so well that each one cannot be picked out

solución
combinación de dos cosas que se mezclan tan bien que no se puede identificar cada una

strategy
a plan that can solve a problem

estrategia
plan que puede resolver un problema

stream
a small flowing body of water that starts with a spring and ends at a river

arroyo
pequeño cuerpo de agua que fluye y nace en una vertiente y termina en un río

survive
to continue to live

sobrevivir
continuar viviendo

— T —

two-dimensional
drawings and sketches that are done on flat paper to show width and height

bidimensional
dibujos y bosquejos que se hacen en papel plano para mostrar el ancho y la altura

V

valley
the low place between two hills or mountains

valle
lugar bajo entre dos colinas o montañas

W

weathering
the breakdown of rocks into smaller pieces called sediment

meteorización
desintegración de rocas en trozos más pequeños llamados sedimento

Index

A

Air 38–43
Analyze Like a Scientist 13, 23–25, 46–47, 55, 58–60, 70–73, 83–85, 95, 113, 117–120, 122–125, 136–139
Animals
    depending on plants 94, 95, 113, 114
    habitat loss and 122–125
    humans' changes to environment and 117–120
    as living things 23–25
    making homes for 136–139
    needs of in habitat 55, 58–60, 103–105
    plants and 89–90
    working with plants 78
Apple trees 89–90
Ask Questions Like a Scientist 10–11, 80–81

B

Bats 94, 95
Beach, litter on 115–116
Bees 87–88, 92–94
Biodiversity
    comparing 34–35
    describing 32–33
Bison 70–73
Bumblebees 87–88
Burrs
    dogs and movement of 80–81, 83–85, 132–135
    hook-and-loop fastener and 86
Butterflies 94, 95

C

Can You Explain? 8, 78
Comparing
    biodiversity 34–35
    habitats 64
Corn 36–37

D

Deserts, life in 20–21, 22
Design Solutions Like a Scientist 4–5, 142–146
Dispersion
    of pollen 96–100
    of seeds 101
Dogs 82, 83–85, 132–135

E

Earthworms 54, 56–57
Environment
    living and nonliving things in 14
    in living landscapes 18–19
Evaluate Like a Scientist 18–19, 34–35, 64, 74, 89–90, 114, 130–132, 140

F

Factories 119
Farmland 119
Fish, climbing 10–12, 66–69
Flowers, bees on 87–88
Food
    habitats and 23–25, 61–63
    as need of animals 19, 103–105
Forest fires 70–73, 119
Forests 61–63, 121. *See also* Trees
Foxes 61–63
Fungi 23–25
Fur 80–81, 82, 83–85, 132–135

G

Grids, sample 25, 26, 27–31

H

Habitat loss 122–125
Hiking, across woody landscape 82
Homes, making for animals 136–139
Hook-and-loop fastener 86
Honey 94
Hummingbirds 94, 95

I

Insects
    bumblebees on flowers 87–88
    helping plants 92–94
    pollination and 110–113
Investigate Like a Scientist 15–17, 27–31, 38–44, 106–109

J

Joshua Tree National Park 22

L

Land cover 103–105
Litter
    on beach 115–116
    as form of pollution 118
Lizards 139

M

Maps 48–51
Math 24–25, 26, 27–31
Modeling
    of pollen spread 96–100
    of seed movement 101, 106–109
Moles 52–54
Mounds, mysterious 52–54

N

National Parks 70–73
Nutrients, as need of plants 46–47

O

Observe Like a Scientist 14, 20–21, 22, 26, 32–33, 36–37, 44, 45, 52–54, 56–57, 61–63, 82, 86, 87–88, 92–94, 101, 102, 103–105, 110–113, 115–116, 121, 126–128, 129

P

Park rangers 70–73
Perch, climbing 10–12, 66–69
Pets 139
Plants
    animals and 89–90
    bees and 92–94
    burrs, fur and 83–85, 132–135
    depending on animals 94, 95, 112, 113, 114
    growth of in different conditions 44
    humans' changes to environment and 117–120
    as living things 23–25
    needs of in habitat 36–37, 46–47
    planting seeds and 45
    working with animals 78
Pollen and pollination
    animals and 95, 112
    bees and 87–88, 94
    defined 94
    insects and 110–113
    modeling 96–100
Pollution
    chemicals as 119
    litter as 118
    preventing 126–128
Primate habitat 137

R

Rain forests 121
Rangers 70–73
Record Evidence Like a Scientist 66–69, 132–135
Restoration 70–73
Runoff, polluted 126–128

S

Sample grids 25, 26, 27–31
Schoolyard
    living things in 13
    mapping sun and shade in 48–51
Seeds
    animals transporting 112
    apple trees and 89–90
    burrs, fur and 83–85, 132–135
    modeling movement of 101, 106–109
    planting of 45
    types of, travel and 102
Shade, mapping of 48–51
Shelter
    earthworms and 56–57
    habitats and 23–25, 61–63
    as need of animals 19, 103–105
    plants and 113
Sleep
    habitats and 23–25
    as need of animals 19
Soil
    living in 52–54
    as need of plants 38–43
    plants' needs and 46–47
Space as need of plants 46–47
Sports, eco-friendly area for 2, 4–5, 142–146
Sunlight
    habitats and 61–63
    mapping of 48–51
    as need of plants 38–43, 46–47

T

Think Like a Scientist 48–51, 96–100
Trees
    apple 89–90
    cutting of 119, 122–125

W

Water
    habitats and 23–25, 61–63
    as need of animals 19, 103–105
    as need of plants 38–43, 46–47
Wind 101, 106–109
Woody landscape, hiking across 82
Worms 54, 56–57

Y

Yellowstone National Park 70–73

Z

Zoos 136–139